P9-APZ-729

i-mode Strategy

Takeshi Natsuno
Managing Director, i-mode Strategy
NTT DoCoMo, Inc.
Japan

Translated by Ruth South McCreery
The Word Works, Ltd.
Yokohama, Japan

WILEY

Translation from the Japanese language edition published by Nikkei BP Planning, Inc.

Telephone (+44) 1243 779777

Email (for orders and customer service enquiries): cs-books@wiley.co.uk
Visit our Home Page on www.wileyeurope.com or www.wiley.com

Other Wiley Editorial Offices

John Wiley & Sons Inc., 111 River Street, Hoboken, NJ 07030, USA

Jossey-Bass, 989 Market Street, San Francisco, CA 94103-1741, USA

Wiley-VCH Verlag GmbH, Boschstr. 12, D-69469 Weinheim, Germany

John Wiley & Sons Australia Ltd, 33 Park Road, Milton, Queensland 4064, Australia

John Wiley & Sons (Asia) Pte Ltd, 2 Clementi Loop #02-01, Jin Xing Distripark, Singapore 129809

John Wiley & Sons Canada Ltd, 22 Worcester Road, Etobicoke, Ontario, Canada M9W 1L1

Wiley also publishes its books in a variety of electronic formats. Some content that appears in print may not be available in electronic books.

British Library Cataloguing in Publication Data

A catalogue record for this book is available from the British Library

ISBN 0470 85101 5

Typeset in 11/13pt Times by Laserwords Private Limited, Chennai, India
Printed and bound in Great Britain by Antony Rowe Ltd, Chippenham, Wiltshire
This book is printed on acid-free paper responsibly manufactured from sustainable forestry in which at least two trees are planted for each one used for paper production.

About the Author

Takeshi Natsuno
Managing Director, i-mode Strategy
NTT DoCoMo, Inc.

Takeshi Natsuno is directly responsible for all of strategy i-mode, the world's largest wireless Internet service, with more than 34 million subscribers. After he graduated from Waseda University, he first joined a leading company in the Japanese energy industry. After gaining extensive experience in real estate development projects there, he entered the Wharton School at the University of Pennsylvania and earned an MBA. Before he joined NTT DoCoMo in 1997 to launch the widely i-mode service, he was an executive vice president at an Internet start-up company from 1996 to 1997. He is well known as the founder of i-mode and was selected as one of the 25 most influential e-business leaders worldwide by *Business Week* in 2001. He has also written *a la i-mode*, a sequel to this volume, published in 2002.

Foreword to the Japanese Edition

At two in the afternoon, Wednesday, August 1, 1997, in the President's office on the tenth floor of the NTT DoCoMo headquarters in Toranomon, Tokyo, the then President, Koji Ohboshi, who is now DoCoMo's chairman, said to me 'We're going to start up a mobile multimedia business based on cell phones!' and handed me a fat report from McKinsey & Company.

My gut reaction was 'This sounds like a cool business.... we have a good chance of success,' but then I realized that the project team had only one member – me. 'What about some staff?' I asked, and Ohboshi replied, 'Bring together whomever you like.' The rest is i-mode history.

As is not unusual for salaried employees in Japanese corporations, my network of connections outside my own company was limited. How on earth was I to put together a team? I called my old friend and business mentor Masafumi Hashimoto, President of a printing company in Kumamoto, and asked him to give me some leads for acquiring the team members I would need to help me.

That is how Mari Matsunaga joined the team. Through her, we also acquired Takeshi Natsuno, the author of this volume. Masaki Kawabata joined us to handle the server side of the business.

It was the characteristics and capabilities of the individual team members combined with our shared conviction that we were bound to succeed that made a success of the i-mode project. To those who want to learn more about the project's early days, I recommend Mari Matsunaga's book, *The i-mode Affair* (*i-modo jiken*, Kadokawa Shoten).

As the number of i-mode subscribers started climbing, the media, even overseas, began to show interest. One day, a television crew from the United States came to do a story on us. The reporter asked, 'Are you interested in history?' When I said yes, he asked, 'Then what historical event do you think i-mode corresponds to?' That was a new way of looking at it. I remember feeling impressed at how different his approach was from the Japanese reporters.

The reply that instantly came to mind was Columbus's discovery of America late in the fifteenth century. Many Europeans agreed with Columbus that the world was round, but he was the one who dared to set out to prove it – and discovered more than he had bargained for.

If the European continent is the personal-computer-based, wired Internet market of today, then i-mode is America, the new world. People had talked about the possibility of Internet access from cellular phones, and some had tried it, but no one had seriously set out to do it – until, that is, our i-mode development team embarked on its voyage of discovery.

Today, five hundred years after that new world was discovered, the Americas have surpassed Europe as a market. i-mode has a similar potential. It will be a market to rank with the wired Internet market. That is why the eyes of the world are on i-mode.

Cellular phones, browsers, a packet-switching network, servers, and content – they make up i-mode. But the technologies (and the content) were already out there. It took no huge invention to make i-mode possible. Many, learning about the process, will think they could have done it too: it is easy when you know how. Why am I reminded of Columbus's egg?

Before launching i-mode, we spent a long, long time hashing it out thoroughly. Where is our market? What is the product concept? What technologies will make it a reality? What about the fee structure? Content? Marketing? How will we drive continuous growth? Those discussions led to the success we see today.

Mari Matsunaga's book is, as it was, a log of the voyage of discovery. Takeshi Natsuno's book describes the seamanship that made it possible.

But this book covers more than the basics of how to navigate. It is a business strategy book that tells the reader how to conquer the incredibly rich new world of mobile multimedia opening out before us.

The world's cellular phone market will change into a mobile multimedia market. Who will be the winners in that market is yet to be decided. Cellular phones, components, network equipment, servers, software, content, and telecommunications providers – enormous opportunities await all those involved in mobile multimedia. I hope that you will read this book, learn from it, and go on to toast your own success in this new world.

So now, let it begin – the story of the start of the i-mode development saga.

October 2000
Keiichi Enoki
NTT DoCoMo Director, Gateway Business

Foreword to the English-language Edition

Two years have already passed since I finished writing *i-mode Strategy* in Japanese, describing the use of the service that has made specially formatted Internet sites and other online content available via NTT DoCoMo's mobile phones. Since then, the number of i-mode users has increased in line with one of my personal theories, which I call the IT Business Principle – namely, that in the case of information-technology business, numbers tend to increase far beyond original expectations. When the book was published, at the end of 2000, there were about 18 million i-mode users in Japan; as of August 2002 the figure had surged to over 34 million. This amounts to more than 80% of DoCoMo's subscribers, and it is more than a quarter of Japan's entire population. (For details of the increase, see Chapter 1.)

For DoCoMo, i-mode has been a major new source of earnings. In the business year ending March 2002, the company's i-mode data transmission (packet transmission) revenues topped ¥700 billion. This consisted of a tremendous accumulation of tiny amounts – ¥0.3 per packet – and accounted for more than 10% of total revenues for fiscal 2001.

Along with the sharp increases in the number of users and volume of revenues, the past two years have seen a dramatic improvement in the content of i-mode services. In January 2001, just after this book was published in Japan, DoCoMo launched its i-αppli service for downloading software applications, allowing users to install additional programs of their own choice on their mobile phones just as they can do on their personal computers. As of August 2002 there were some 15 million i-αppli users; in other words, close to half of all i-mode subscribers were using this additional software. The programs are compiled using the Java programming language developed by Sun Microsystems; the 15 million i-αppli users can be seen as constituting the largest Java community in the world.

Since the second half of 2001, DoCoMo has been working to extend the range of 'scenes' where mobile phones can be used. The range of additional uses that are now turning into reality include buying drinks from vending machines, making purchases at convenience stores, and withdrawing cash from bank ATMs. We are approaching the time when mobile phones will become 'electronic wallets', as I advocated from the beginning.

The new high-speed data transmission service FOMA (Freedom Of Mobile multimedia Access), launched in May 2001, is making i-mode even more attractive by allowing users to send and receive large volumes of data more quickly and easily than before. This makes it possible, for example, to download more complicated software applications. And with 'i-motion' users can download and watch video clips. I see this as a field with great potential for further development of i-mode services.

Another change over the past two years has been the internationalization of i-mode. DoCoMo has entered into alliances with overseas telecommunication carriers that recognize the merit of the i-mode business model, including KPN Mobile (Netherlands) and AT&T Wireless (United States), and i-mode gradually is becoming available in other countries. In my own work, I am constantly traveling to other Asian countries, Europe, and America to promote the further spread of i-mode services around the world, as well as striving for the further improvement of the services available in Japan.

In this book I offer an extensive introduction both to the basic i-mode concept, which is grounded in the theory of complex systems, and to the actual operation of i-mode services. I hope that it will both contribute to an understanding of i-mode and offer hints for readers in developing their own IT businesses.

Takeshi Natsuno
August 2002

Contents

Color Plates

Plate 1. 'I never dreamed we'd be on the cover of *Business Week* this soon.' *Business Week*, January 17, 2000. Copyright The McGraw-Hill Companies, 2000.

Plate 2. The Content Portfolio.

Plate 3. Magazine advertisement.

Newspaper advertisement.

Marketing i-mode in a way consumers would understand. The advertisements used no IT terminology and stressed convenience for the user.

In-train hanging advertisement.

Over 100 i-mode books. (See Chapter 5, Section 5.11 for details.)

Plate 4. The 501i Series.

The 502i Series. © Tetsuka Production. © Sotsu Agency, Sunrise.

Plate 5. The 209i Series.

The 821i Series.

Plate 6. The 503i Series.

Chapter 1
Success

1.1 IT Businesses Grow far More Than Expected or do not Grow at All

Information Technology (IT) businesses grow far more than expected – or they do not grow at all. Never do they grow little by little along predictable lines. In terms of that IT law, i-mode is clearly in the 'grow-far-more-than-expected' category.

NTT DoCoMo launched i-mode service on February 22, 1999, barely a month after we unveiled the new service to the media at a press conference in Tokyo (Figure 1.1) on January 25. We had chosen the actress Hirosue Ryoko to appear in the commercials, and inviting her to the event helped draw a good crowd, with about 500 members of the press present.

The event began with Keiichi Enoki, then manager of our Gateway Business and now DoCoMo director, announcing our new service and new types of cellular phones. He was the person with overall responsibility for i-mode. When asked by reporters what the target figure was for the number of subscribers to i-mode, Enoki replied, 'We hope to sign up between two and three million in the first year of service. Our goal is 10 million subscribers after three years.'

1.1.1 The Pace Outstripped Our Expectations

The results far exceeded Enoki's prediction. We hit our initial target of one million subscribers on August 8, 1999, six months from the start of the service. By way of comparison, it had taken 13 years for DoCoMo to reach

Figure 1.1 The i-mode launch party. (Photograph: Takanari Yagyu.)

the one million mark with mobile phone subscribers. We regarded the one million mark as a crucial milepost, a subscriber base giving us critical mass. Once that level was passed, we would be into a positive feedback cycle in which people would notice what others around them were doing with their i-mode phones and would want to use the service themselves.

As we had expected, once the number of subscribers passed one million, the pace at which we added new subscribers surged. Adding the second million subscribers took only two months.

It then took just two additional months to reach our first-year target of three million subscribers, on December 23, ten months after the service began on February 22. We were three months ahead of our own projections.

In 2000, the rate at which we gained subscribers accelerated. In the first half of that year, it took only six weeks, on average, to add another million i-mode users. The numbers climbed steadily past four million, five million, and broke through the 10-million level on August 6, 2000. In under 18 months, i-mode had become a huge hit, with over ten million subscribers.

1.1.1.1 One Million in 20 Days: 50,000 New Subscribers a Day

At that point, we were adding a million new i-mode subscribers every three weeks – a pace set in July 2000, and sustained ever since. (Figure 1.2). Our average daily increase was 40,000 to 50,000 subscribers. At this point of writing (late October 2000), there are over 14 million i-mode subscribers, and we are seeing no loss in momentum. In fact, the rate at which people are signing up for i-mode is rising.

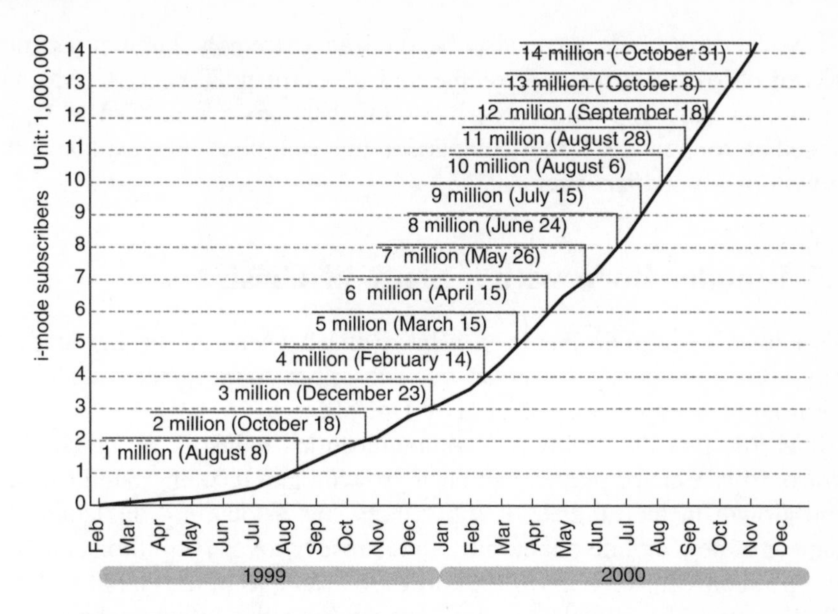

Figure 1.2 i-mode subscribers: rapid growth. *Source*: NTT DoCoMo.

After we passed the 10-million mark on August 6, we surged on to 11 million on August 28: a gain of one million in only 22 days. We reached 12 million in another 21 days, on September 18, and 13 million in 20 more days, on October 8.

1.1.1.2 One-Third of All Internet Users

We at NTT DoCoMo had not dreamed that the drive to sign up for i-mode would be so powerful. At his regular press conferences, Keiji Tachikawa, president of DoCoMo, was kept busy issuing upward revisions of the number of subscribers predicted by the end of fiscal 2000 (the end of March 2001). Our plan was to have 11 million subscribers by the end of that fiscal year, but it now appears virtually certain that we will sail past the 17-million mark. Our next target is 20 million.

Looking at the figures of the number of people having Internet access in Japan, we realize how astonishingly large was the 10-million subscriber base we had by August 2000. According to the Internet Association of Japan's *Internet White Paper 2000,* by the end of fiscal 1999, Japan had slightly fewer than 20 million Internet users. The subsequent growth in i-mode subscribers pushed that number far higher, so that now, by our count, one-third of all Internet users in Japan access the Net via i-mode.

i-mode subscribers have also become an extremely large proportion of DoCoMo subscribers. At of the end of October 2000, DoCoMo had 33 million mobile phone subscribers. Of them, over one-third had also signed up for i-mode. Either comparison makes it clear how explosive the growth of i-mode has been.

1.2 i-mode: Born with a Sense of Crisis

We know that i-mode has been a hit. But what was the background? It may seem odd to say so, given our subscriber figures, but i-mode was born out of a sense of crisis at NTT DoCoMo.

It is true that the mobile telecommunications industry, in which NTT DoCoMo is a major player, had been growing at a dazzling rate in Japan. The growth in the number of subscribers was astounding. In Japan, the number of new mobile phone subscribers rose by over 10 million annually for three consecutive years – 1996 to 1998 – and, as of the end of October 2000, totaled 56 million, according to a study by the Telecommunications Carriers Association.

The number of subscribers was not the only way the industry was growing: market scale was also expanding rapidly. Total sales by providers of mobile telecommunications services stood at ¥1,407,400 million in fiscal 1995 (April 1995 through March 1996) and had mushroomed to ¥5,207,800 million in fiscal 1999. The market as a whole had grown 3.7-fold in only four years.

During those years of explosive growth, DoCoMo was also growing at a highly satisfactory rate. The number of subscribers had rocketed from 2,210,000 in fiscal 1995 to 23,000,000 in fiscal 1999 – a more than tenfold increase.

NTT DoCoMo's operating revenues reached ¥3,718,700 million in fiscal 1999, rivaling those of our parent company, NTT. And the company was profitable as well: the DoCoMo Group was generating about 60%, or ¥500 billion, of the recurring profit of the NTT Group as a whole.

1.2.1 A Multidisciplinary Team

If the company was performing so strongly, where was that sense of crisis coming from?

The in-house project that led to the launch of i-mode service dates back to January 1997. The starting point was when NTT DoCoMo's president

Koji Ohboshi (now chairman) appointed Keiichi Enoki as project leader and told him to start a new mobile phone service. That was the year the now-famous author of the *i-mode Incident*, Mari Matsunaga, formerly editor in chief of *Torabayu*, a job-search magazine for women, was head-hunted from Recruit. That same year, I myself was invited by Matsunaga to participate in the project.

The DoCoMo Gateway Business division, which is in charge of i-mode, was a hodgepodge of people from various backgrounds. Project leader Enoki came to us from NTT. His career there began when it was still the state telephone monopoly, Nippon Telegraph and Telephone Public Corporation; before joining our project, he had been the NTT branch manager for Tochigi Prefecture. His background was in engineering.

Using DoCoMo's first in-house call for applications, Enoki gathered a group of young DoCoMo employees to form the new project team, with the addition of people like Matsunaga and myself, whom he brought in from outside, and representatives from the equipment manufacturers. We were a very diverse group.

It is common practice for private sector companies to hire from outside and to open up a new operation for those wanting in-house transfers. But for DoCoMo, with its rather bureaucratic roots, those were highly unorthodox ways to recruit staff. But necessity was the mother of invention: when put in charge of what became the i-mode project team, Enoki was its sole member, and he was given a free hand to assemble the people he needed. The resulting diversity of skills and attitudes was a major factor in i-mode's success.

1.2.2 Towards the Second S-Curve

But, I digress.

About the time he was launching this in-house project, Ohboshi had pointed out that DoCoMo's growth curve could be heading into a second S-curve. He had prepared a chart (Figure 1.3) delineating the two S-curves for a management policy document entitled 'Shifting from Volume to Value,' released in July 1996.

The year 1996, when that report was written, was the first of three years in which mobile phone services acquired 10 million new subscribers annually. It was the first year of explosive growth in the mobile telecommunications industry. It marked the start of the first S-curve, when the number of subscribers began surging upward. Those were the years when

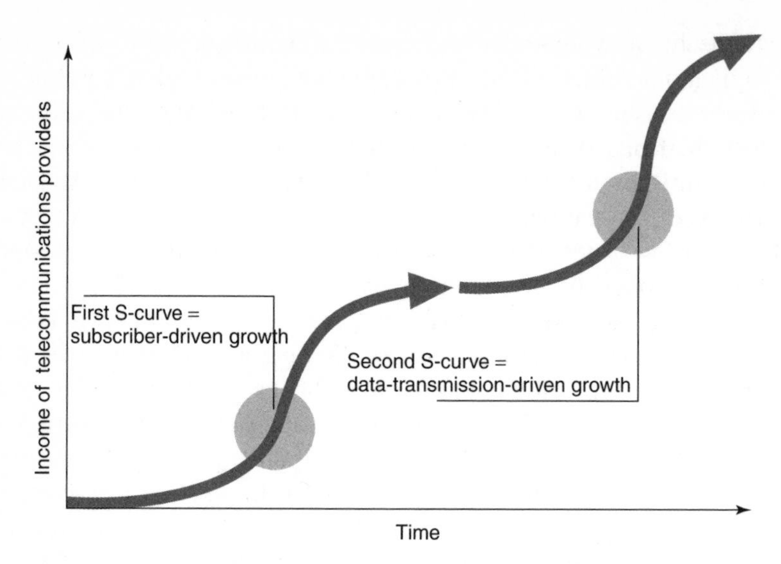

Figure 1.3 Toward the second S-curve.

operating revenues from transmission fees grew enormously for companies in our industry. That was the first S-curve, the growth curve due to volume.

It was obvious, however, that that rate of growth in the number of subscribers could not be sustained. Once all the consumers who wanted to use mobile phones had signed up, that growth would be over.

At that time, the total number of mobile phone subscribers was soon to reach 40 million for Japan as a whole. While the dominant view now is that the market will grow to about 80 million, at that time, 60 million was widely thought to be reasonable, extrapolating from the approximately 60 million fixed-line phone subscribers.

With 40 million subscribers already signed up, the assumed possible growth was another 20 million, before hitting the upper limit of 60 million. These were figures that weighed on telecommunications providers' minds.

1.2.2.1 From Volume to Value

If the growth rate flattened, telecommunications providers' capacity to generate revenues would decline. In the zero-sum game that would be played out if the number of subscribers was constant, competition between service providers would drive down rates. The fixed-line telephone business in the United States provides a prime example of what happens when heated competition goes after a limited pool of subscribers: one month free for changing phone companies, rebates, and other discount schemes that cut into revenues. The result is a war of attrition that erodes the strength of all the participants and potentially leads not only to a decline in capacity to develop new services but also to a risk of a fall in the level of service provided by the industry as a whole.

Ohboshi offered an alternative, in an easy-to-understand way, by creating the concept of the second S-curve. That is, instead of devolving into a zero-sum war over a limited pool of subscribers, we could create a new market, data communications, in addition to voice communications, and thereby move into another growth phase.

In this new growth phase, revenues could continue to grow even if the number of subscribers did not, because subscribers would be using their mobile phones for purposes other than voice communication. That is the second S-curve, the 'value' growth curve.

Riding the second S-curve would require pioneering a service that would increase subscribers' use of their mobile phones, a new service that would generate data traffic. i-mode was to become that service.

1.3 Evolution of a Text-Based e-Mail Culture

Back in 1996 we were already seeing signs of growth in data traffic on mobile phones. Some subscribers, particularly young people who had earlier become adept at using pagers, were attracted to e-mail services using mobile phones, called *text mail* or *text communication*. Young women of high-school, or college-going, age could be spotted riding the commuter trains or walking down the street, staring at their mobile phone screens and manipulating their ten-key pads.

That young user segment was creating a new telephone culture. First, between 1994 and 1996, pager usage soared, only to fall just as swiftly between 1996 and 1999. With pagers, text messages were a one-way street, from phones to pagers. Those were the days of 'one-way service' (Figure 1.4).

1.3.1 From One-Way to Interactive

The decline in pager use did not signal an end to the new telephone culture. Instead, from around 1997, use of e-mail services on cellular and Personal

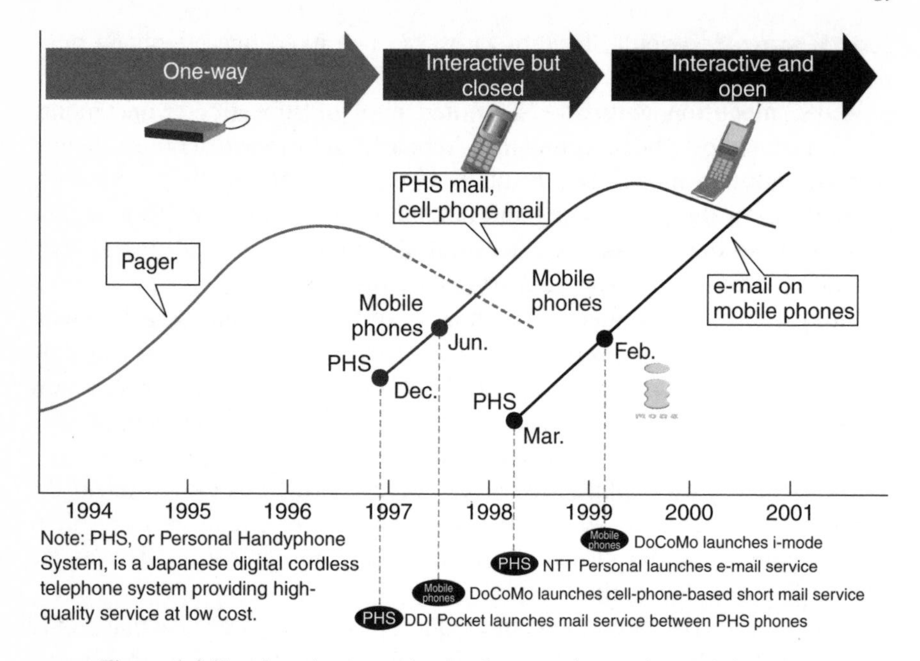

Figure 1.4 Text-based communication becomes interactive and then open.

Handyphone System (PHS) phones began to increase. (The DoCoMo service was called *Short Mail*.) The generation that had deftly tapped out messages on their telephones to send to their friends' pagers gradually matured into cellular and PHS phone users. At that stage, interactive exchanges of messages were possible – but only between cell phones or between PHS phones and only between phones with the same provider. The systems were, thus, interactive but closed.

Then came the age of interactive, open e-mail using mobile phones. Now it was possible to send and receive e-mail messages between mobile phones operating on different providers' networks and also between mobile phones and personal computers with Internet access. Over the Internet, e-mail could now be exchanged with any other Internet-capable terminal.

It was just at the dawn of this interactive, open e-mail era that the work of developing i-mode began. We had no doubt that interactive, open services would flourish. Our thinking focused, in developing i-mode, on how to provide the optimal context – ease of use and attractive fee structures, for example – to enable interactive, open e-mail to realize its potential.

1.4 Beyond Talk

The substance of DoCoMo's policy of 'making the value shift' and riding the second S-curve was to create new situations in which people would use their mobile phones.

When we launched i-mode, the company public relations department created the line 'beyond mobile phones to talk on to mobile phones to use'. It was quite apt, for we had prepared a menu of a variety of services to be used by as wide a range of age groups as possible, not just the young trendsetting segment. We called that our content portfolio. That portfolio of services was highly diversified: subscribers could check their bank balances, make bank transfers, reserve airline or other tickets, check restaurant guides, research train connections, use online dictionaries, shop online, and even play games online. (See the color plates showing examples of content at the beginning of this book.)

Of course, no one subscriber would be likely to use all these services. We were aiming for an overall balance and a wide range of offerings so that people who differed widely in age group, gender, interests, and preferences would find services tailored to their needs.

DoCoMo did not develop those services by itself but in conjunction with the service providers. Those jointly developed services are what are commonly called the official i-mode sites. At the outset, we had a menu of official sites provided by 67 developers. Their number increased rapidly, so that as of October 2000, we have about 1200 official menu sites offering services from 665 companies. The DoCoMo official sites menu begins from what is called the iMenu, the top page for i-mode.

1.4.1 Content-Packed: The Virtuous Cycle

Apart from sites listed on the official menu, a huge number of other Internet sites have been set up to be accessed with i-mode. These are what we call the voluntary sites, also known as the 'do your own thing sites'. Subscribers can access them from their mobile phones by inputting the URL. As of the end of October 2000, they totaled about 28,000 sites (according to Oh! New?, an i-mode directory and search service provided by Digital Street). DoCoMo does not manage those sites, as may have been obvious by my quoting another source for the number of such sites. As is the case for Web sites on the Internet as a whole, trying to make an accurate count is hopeless; we in fact never intended to. Our notion is that businesses or individuals are free to think up whatever services they like and provide them via i-mode.

Between the official sites on the iMenu and the voluntary sites, a wealth of content is available by i-mode – and those riches are the drivers attracting more and more new subscribers to i-mode. The more subscribers we have, the more the service menu grows, and the more the service menu grows, the more we attract users to see what is there. The ongoing positive feedback cycles are why i-mode has been a hit with over 14 million subscribers (Figure 1.5). Plenty of content is the key to the i-mode growth cycle.

1.4.2 Active Users: 95%

The story is not merely that there are so many i-mode subscribers: most use it frequently; i-mode is now part of their lives. Several types of data indicate how deeply rooted i-mode has become. First, usage rates are extremely high. Looking at the figures as of September 2000, we find that of 12,330,000 subscribers, about 87%, had accessed the Web from their i-mode phones at least once in the last week of September (Figure 1.6). In the same week, about 78% had used their phones for e-mail.

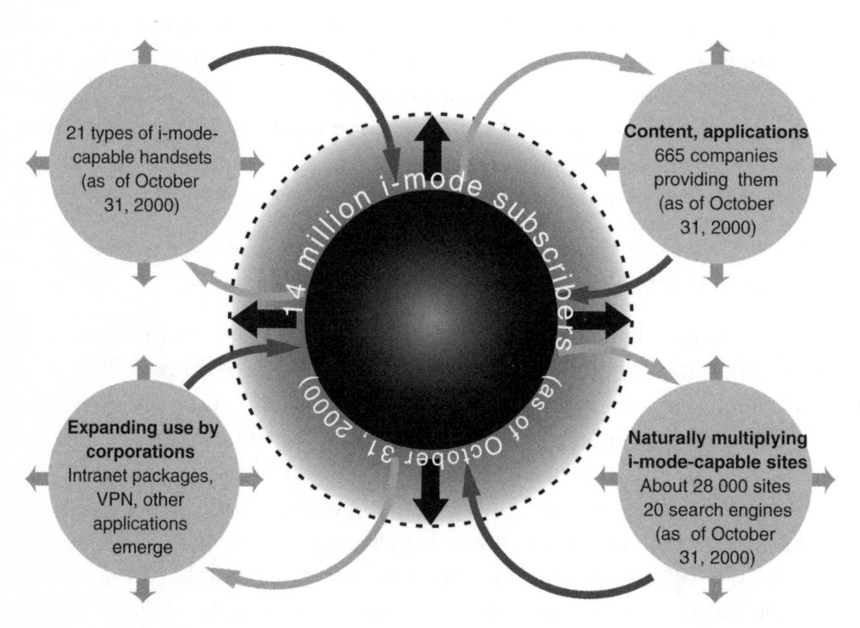

Figure 1.5 Positive feedback for i-mode. *Source*: DoCoMo and Digital Street, Inc.

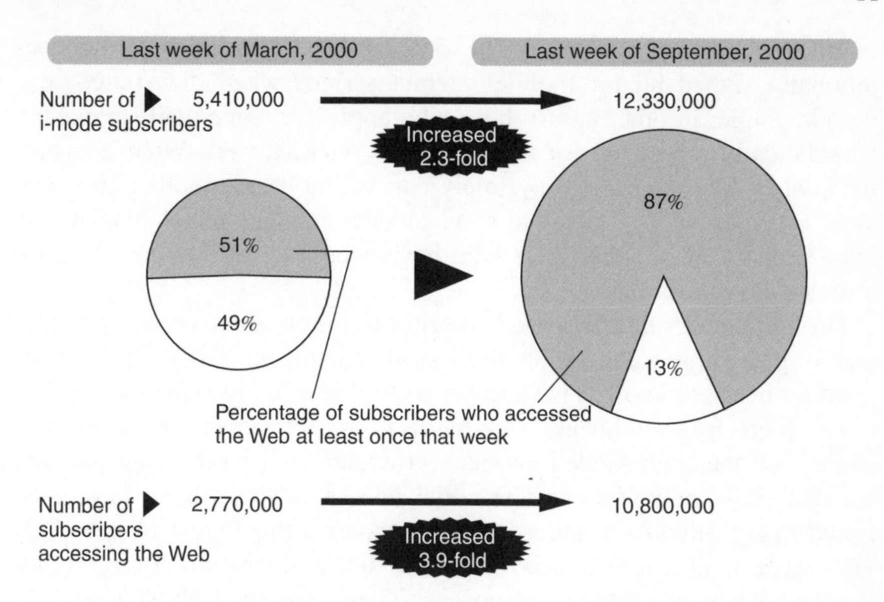

Figure 1.6 Usage rate for i-mode Web access service. *Source*: DoCoMo.

Total usage is 80 million e-mail messages transmitted via i-mode – per day. That works out to an average of seven messages a day sent or received by each i-mode subscriber.

A survey done in June 2000 indicated that off all i-mode subscribers, only 5% had never used its functions – subscribers whose bills for packet usage were zero. That is a remarkably low figure. While I have seen very few instances of other companies making public similar figures on usage rates by subscribers, I would be very surprised if any other mobile communications company had such a low nonusage rate.

1.4.3 i-mode in Mind When They Sign Up

Why is it that so few of our subscribers fail to make use of the i-mode functions? The answer is simple: we are attracting subscribers who choose us because they very much want to use the services available on i-mode. The monthly usage fee for i-mode is ¥300. When someone purchases a mobile phone, the salesperson must always say, 'There is a ¥300-a-month usage fee. Do you want to sign up for i-mode?' Only those customers who say yes can use i-mode services; those without any interest in using them do not sign up.

What about other companies? In some cases, all subscribers are unconditionally signed up for mobile Internet service when they purchase a mobile phone; in other cases, those who apply for voice mail service are automatically signed up for the Internet service as well. What happens, then, when Internet service is simply part of the basic mobile phone service or comes with an optional service, such as voice mail? A relatively large proportion of subscribers who have access to the Internet via their mobiles never use that service.

The number of subscribers is, by itself, meaningless to a service provider that is providing value-added services to customers. What that service provider needs to know is the number of people actually using the service.

The high usage frequency of i-mode faithfully reflects the usage frequency of the value-added services provided on i-mode. Consider the Tsutaya chain of video and CD rental and sales shops: they have used i-mode very effectively indeed. They announce the arrival of new CDs via i-mode mail and also issue electronic discount coupons. Both promotions evoke huge customer responses. Those informed about new CDs by i-mode mail outbuy others by a large proportion. Those who receive electronic coupons have total monthly rental fees that are 59% higher than the other members. The effects are clear (Figure 1.7).

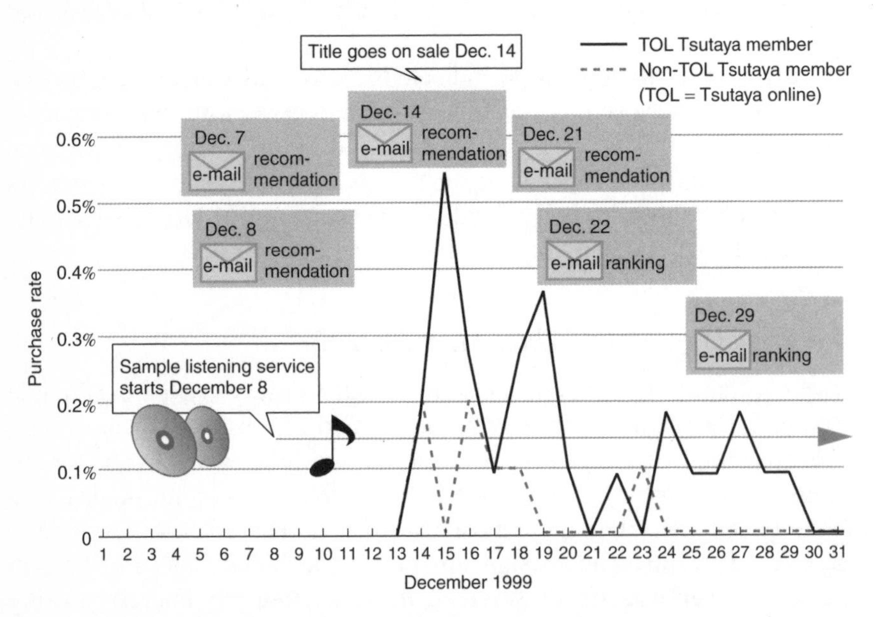

Figure 1.7 CD purchase rates at Tsutaya. *Source*: Tsutaya Online.

The airlines have also done well by supporting ticket purchase by i-mode: each airline using i-mode has sold tens of millions of yens worth of tickets via i-mode each month. E-commerce sites selling computer games are also said to be racking up monthly sales in tens of millions of yen via i-mode. DLJdirect SFG Securities, which deals in stocks by phone and e-mail, reports that it receives about 20% of its business by i-mode.

1.4.4 Information Close to Home Works

An interesting aspect of i-mode is that the proportion of women using it is higher than for the wired Internet. As of the end of October 2000, 41.7% of i-mode users were women. Recently, we have seen a steady increase in sites appealing to women, and, with increasing coverage in women's magazines, we expect that the proportion of women among our subscribers will continue to rise.

Many seem to think that i-mode is a medium for the young, that the main demographic group using it is people in their twenties. It is true that i-mode has made great strides in that age group, which accounts for 43% of our subscribers. But older age groups are also proving to be highly receptive to i-mode. The over-40s age group accounts for 29% of all i-mode subscribers, and that percentage continues to rise. Popular i-mode services in that age group are business news, stock trading, and sports news. Here we see the effect of our broad range of menu options (Figure 1.8).

DLJdirect SFG Securities' analysis of i-mode usage trends by age group has produced some fascinating results. This online securities company reports that the older the age group, the more likely it is that its members are buying and selling stocks by i-mode. While on average 20% of customers of all age groups are trading by i-mode, among those aged 60 and above, the proportion rises to nearly 30% (Figure 1.9).

For many individuals, i-mode provides services that they cannot live without, services that fit with their lifestyles. And i-mode, like the Internet itself, offers specialized information services and publications for small communities that are difficult for the mass media, whether television or print, to reach. Other services are also developing to make use of i-mode's ability to reach strictly targeted communities.

For example, Surflegend, a surfer site, provides information on the state of the surf all around the Japanese coastline. This site, which already has over 100,000 subscribers, enjoys the enthusiastic support of surfers

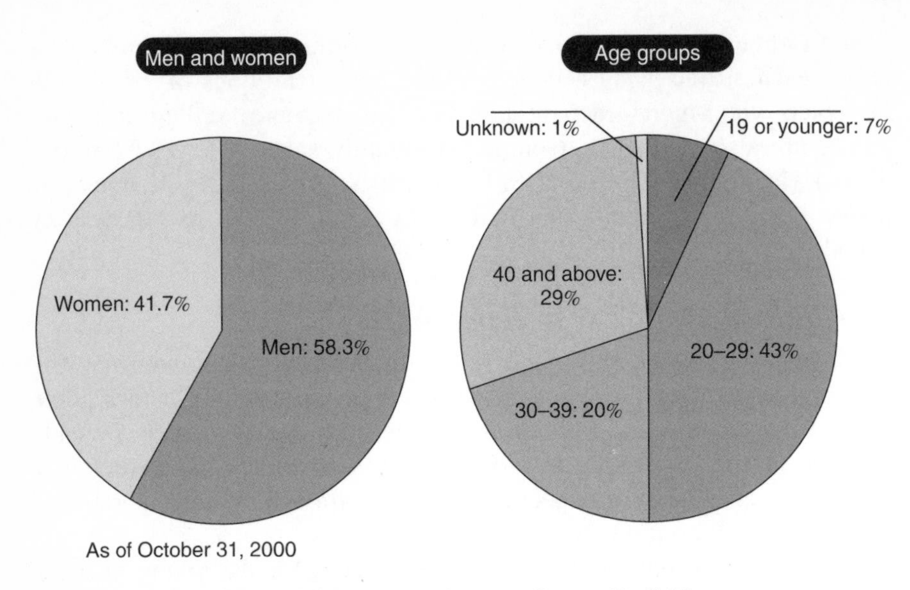

Figure 1.8 Mode–mode users. *Source*: DoCoMo.

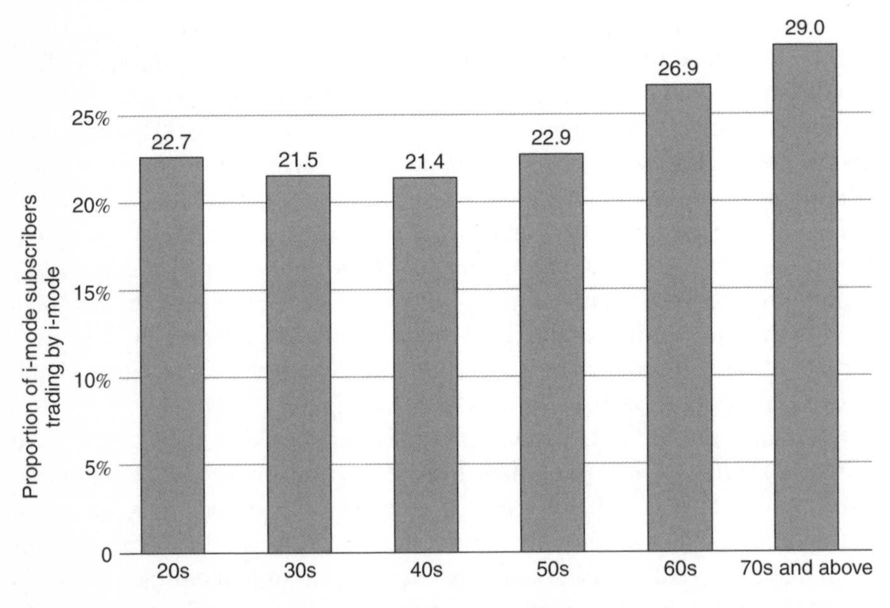

Figure 1.9 Proportion, by age group, of DLJdirect SFG Securities' customers trading by i-mode. *Source*: DLJdirect SFG Securities.

throughout the country. They check the information on Surflegend and then decide, 'The waves are going to be good over there today – why not head there this afternoon?' Because it is difficult to get such information in real time, and about all the many areas of our long coastline, surfers are willing to pay for this service.

Surflegend, by the way, is based on a nationwide network of surf shops. Because knowing the state of the waves is important to such shops, they usually have their employees checking out the surf at about 4 a.m. each day. The information they collect is brought together on their network – and the result is Surflegend. Surf shops all around the country collaborate to provide the information that Surflegend passes on to subscribers.

1.4.5 Average Monthly Usage Fee is over ¥2000

As these examples indicate, i-mode makes life more convenient and has thus become an essential part of many users' lives. Naturally, i-mode also has had extremely large effects on DoCoMo. These effects are of three types.

1.4.5.1 Increase in DoCoMo's Operating Revenues

The first effect is an increase in DoCoMo's operating revenues (Figure 1.10).

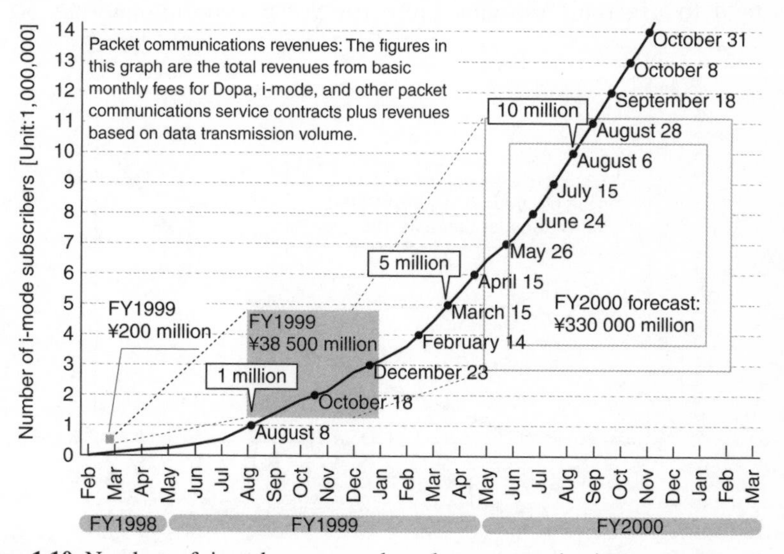

Figure 1.10 Number of i-mode users and packet communications revenues. *Source*: DoCoMo.

In fiscal 1998, when we launched i-mode, DoCoMo had only ¥200 million in revenues from packet communications (i.e. from nonvoice or data transmission). The fiscal 1998 figure included only slightly over a month's revenue from i-mode; income from other forms of packet communications was almost nil.

In fiscal 1999, the first year for which we have a full 12 months' contribution by i-mode, DoCoMo's packet communications revenue surged upward to ¥38,500 million. Naturally enough, in fiscal 2000, with the even stronger continued growth in the number of subscribers, even larger revenues were virtually guaranteed. In September 2000, for example, monthly packet communications charges were averaging over ¥2000 per subscriber. Not wanting to cause our subscribers to go bankrupt through overuse of i-mode, we ask our content providers to design their sites so that they do not take too long to load, but even so, the average packet communications charge per subscriber is continuing to grow (Figure 1.11). With 14 million i-mode subscribers each spending, on an average, ¥2000 per month on packet communications, DoCoMo's packet communications revenues total ¥28 billion per month.

1.4.5.2 Internet Services That Need a Mobile Phone

Another interesting revenue trend for DoCoMo is that i-mode subscribers also tend to use their mobiles more for voice communications, so that

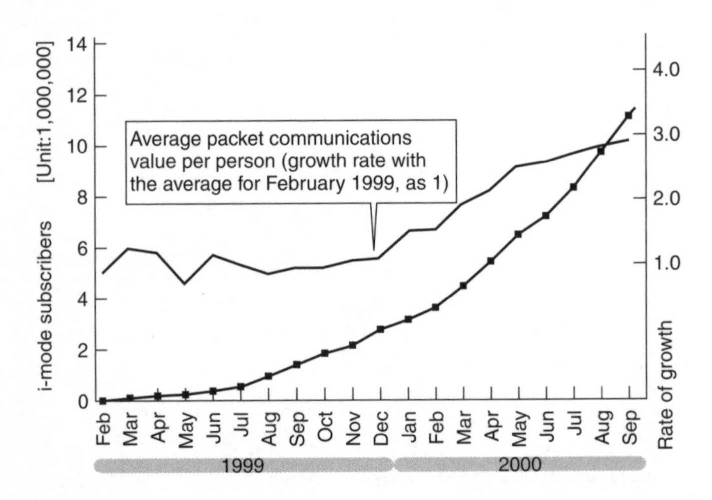

Figure 1.11 Growth in average packet communications volume per person and in number of i-mode subscribers. *Source*: DoCoMo.

average revenue per user (ARPU) is higher for both packet and voice communications. On the basis of a sample of our subscribers, we found that i-mode subscribers make about 20% more mobile phone calls than do DoCoMo subscribers without i-mode. This is partly because screens in i-mode have a 'phone to' function – all the user has to do is click on a phone number to dial it. This is a very handy feature. If, for example, you find a restaurant that looks good on an i-mode restaurant guide, you can just click on the phone number shown on the screen to call and make a reservation. This is not possible using the Internet from a personal computer. It takes a mobile phone – a device that combines Internet access with voice communication functions.

Another revenue source for DoCoMo comes from information service fees, also called content access charges; of the ¥100–¥300 that subscribers pay content providers for access to those services, 9% is retained by DoCoMo as a handling fee for acting as their agent in collecting the fees. Subscribers who use the information services on the official i-mode menu, which charges a fee for access, receive one bill from DoCoMo for their communications charges and content access charges, and DoCoMo retains part of the content access charge as its handling fee. The handling fees are peanuts compared with the packet communications charges, but still add up to over ¥200 million a month.

Another effect that i-mode has brought about is laying the groundwork for the spread of new mobile communications services known as 3G (third-generation) or IMT-2000. DoCoMo plans to start offering 3G services in late May 2001. What is special about 3G mobile communications services is their faster transmission speeds: 64–384 kbits/s. DoCoMo will make a huge investment to build the network for this service throughout the country. If ways to use these high-speed transmission capabilities effectively are not forthcoming, DoCoMo will be unable to recover that investment.

Here again, i-mode makes the difference. We will use our new 3G network as the infrastructure for a new, high-performance i-mode service that can transmit full-motion video and large volumes of data. We anticipate being able to attract customers seeking higher value-added services to the new infrastructure. Our i-mode subscribers will take naturally to the new full-motion service, for example. Services already exist that display a series of downloaded still images to show simple animated clips. They have proved quite popular, and people who use these services will want images with more natural motion and therefore migrate to 3G.

Many subscribers are quite conservative when it comes to new technologies and services and might hesitate if suddenly told 'you can see moving pictures on your mobile phone', but we do not have to worry about that – i-mode has prepared the ground for transmitting full-motion images.

1.4.5.3 Preventing Churning

The third effect of i-mode has been a large increase in the number of DoCoMo subscribers. We have not only maintained our market share in mobile communications but also increased it significantly.

In the year and a half after i-mode service began (February 1999 to August 2000), DoCoMo attracted 8 million new subscribers, boosting our total from 24 million to 32 million. Because the number of mobile phone subscribers was then increasing by nine to ten million annually in Japan, our growth may seem unsurprising. DoCoMo had, however, been losing ground in the extraordinarily intense competition between mobile phone companies when we launched i-mode, in the spring of 1999.

Back then, DoCoMo's voice quality was nothing to boast of, especially in comparison with what the mobile phone companies that were part of other telecommunications groups were offering. An earlier jump in the number of DoCoMo subscribers had produced a relatively large number of subscribers for the bandwidth we had available. From 1998 on, the J-Phone group had been skillfully utilizing that weakness to expand their market share. They stressed the high quality of their voice communication service, with advertising copy saying 'You're rarely disconnected' or 'The sound is great!' Their youth-oriented marketing program was also working, and they were steadily increasing market share.

IDO and the DDI Cellular Group (now merged as the KDDI Group) were employing the same strategy, with the cdmaOne cellular service. Having adopted cdmaOne, which was then a new telecommunications technology developed in North America, they were eating away DoCoMo's market share. Acting together to fill the gaps in each other's service areas, they rolled out cdmaOne nationwide starting on April 14, 1999. I remember very well the aggressiveness with which Tsukuda Takeo, who was president of IDO at the time, announced, 'cdmaOne will take away market share from DoCoMo'.

1.4.6 Market Share Bounces Back

They were right. In May 1999, when cdmaOne service became fully available nationwide, DoCoMo's market share fell precipitately. That month,

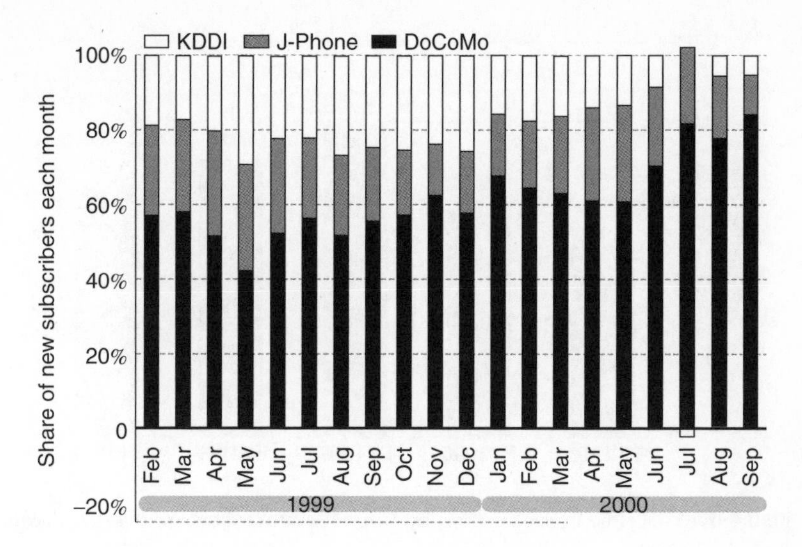

Figure 1.12 Share of new subscribers each month hit 80%. *Source*: Telecommunications Carriers Association.

DoCoMo continued to attract more new subscribers than our competitors did, but our share of new subscribers fell from 49.9% the previous month to 41.8%.

Then, however, as the popularity of i-mode grew, DoCoMo struck back. We stunned the market by capturing 84% of new mobile communications subscribers in July 2000, followed by 78% in August. In 1999, we were attracting 40%–60% of new subscribers each month; in 2000, thanks to i-mode, that figure was ranging between 60% and 80% (Figure 1.12).

Churning – subscriber cancelations – had also been a growing problem for DoCoMo, but that trend reversed as well (Figure 1.13). Other factors such as discounts for full-year contracts, bundled contracts, and other pricing policies helped to limit churn, but i-mode's popularity made a considerable difference as well.

1.5 New Services Emerge

We at DoCoMo are not the only ones to profit from i-mode. Those providing information services on i-mode also benefit, in at least four ways.

One is that content providers receive income. Examples include companies providing news services, ringtone melodies to download, or games:

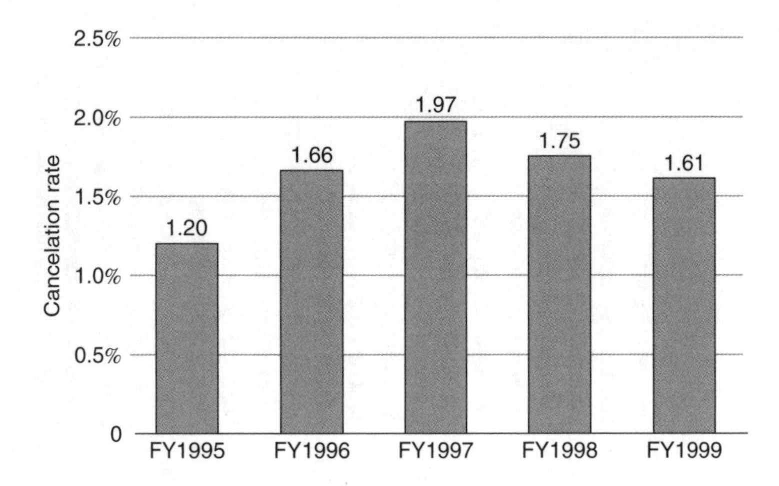

Figure 1.13 Decline in cancelations by DoCoMo subscribers. *Source*: DoCoMo.

these are fee-based services, and each sets its fee at between ¥100 and ¥300 a month. Such fee-based services make up about 30% of the official i-mode menu. For the companies offering such services, providing content via i-mode has been a new revenue stream.

For example, Forever Kyarappa, a popular service provided by the game maker Bandai, has, as of September 2000, about 1.4 million subscribers. This service sends a cartoon character image via the mobile phone network to the subscriber's mobile phone everyday. The subscriber can register the cartoon as the default screen image – the wallpaper – on the mobile phone.

Bandai offers Forever Kyarappa for ¥100 a month; between it and the game content, the company earns, in total, over ¥200 million as revenue from i-mode in a month, that is, ¥2.4 billion annually. Recently, Bandai set up a separate company, Bandai Networks, to specialize in providing Kyarappa and other content over the cellular networks.

The news sites offering fee-based services have not attracted as many subscribers as Kyarappa has. The Nihon Keizai Shimbun, the leading business daily, has about 90,000 subscribers paying ¥300 a month. A major newspaper charging ¥100 a month has about 300,000 subscribers, bringing it about ¥30 million in monthly revenues. That is far less than the newspaper's print run and sales, but quite comparable with business or special-interest magazines, which usually have print runs in the tens to hundreds of thousands.

Not all the providers of fee-based services are large corporations with existing stores of content. Small start-up firms also have actively entered this field because the barriers to creating content for i-mode are much lower than those for television or even newspapers and magazines. Their mobility and dynamism enable them to detect niche markets that the big players overlook, and they provide information services precisely keyed to them.

1.6 A Business Model the Wired Internet Cannot Support

Until i-mode, it was extremely difficult to generate a revenue stream – much less a profit – by providing information over the Internet. Attracting subscribers was difficult. Several thousand subscribers seemed to be about the limit; even highly popular 'adult' sites were barely able to scrape together 30,000 subscribers. To serve those small customer bases, companies charging money to access sites have to have their own servers and databases to identify subscribers and store customer information. That is extremely inefficient. It would not be surprising if quite a few companies with good ideas for such services gave up on delivering them via the wired Internet.

But DoCoMo takes care of the billing and collection for the content provider – in our business model we provide that platform. When we at DoCoMo told content providers that we would collect the fees and that after we deduct our commission, over 90% of the fees would flow into the content providers' pockets, the response was astonishing. Companies that had been able to earn almost nothing by providing information on the Internet leaped at this new opportunity.

Naturally enough, DoCoMo wants the various fee-based content providers for whom we handle the billing to improve the quality of their services. We cannot have them simply make available on i-mode the same content they provide on the Internet because that would not enable us to attract more subscribers with the promise of great services – and thus to be able to collect more commissions for handling the billing.

Hordes of content providers want to be placed on DoCoMo's official i-mode menu, but we tell them that to be listed there, they have to provide sufficient value to satisfy i-mode subscribers. We are happy to expand the offerings on the official menu, but only if the new services meet our requirements.

1.6.1 Birth of a ¥30 Billion Annual Market for Fee-Based Services

How big is the market for fee-based services on i-mode? There are two important indicators. One is the proportion of our subscribers who have signed up for paid services, a measure that focuses on the number of our subscribers who have signed up for at least one such service. Another is growth in the total number of contracts for such services. That indicator would count one subscriber who has signed up for three fee-based services three times.

Looking at the first indicator, we find that as of the end of October 2000, when there were 14 million i-mode subscribers, 48% had signed up for fee-based content services. That is, nearly 7 million people were receiving at least one fee-based service via i-mode. This nearly 50% ratio is astounding in comparison with the proportion of wired Internet users who access fee-based services. In the wired Internet world, ISPs that have actually bothered to set up a fee-based menu of services are lucky to attract 1% of their customers to it. Such services tend to top out with a few hundred to a few thousand subscribers. With i-mode, however, the ratio of subscribers signing up for fee-based services keeps rising.

The number of contracts to receive fee-based i-mode services per subscriber also keeps growing. In August 1999, when we had about 800,000 i-mode subscribers, subscribers had signed contracts for 300,000 fee-based services. That is, total contracts for fee-based services equaled 37% of the figure for total subscribers. In September 2000, when the number of i-mode subscribers broke through the 10-million mark, the ratio of contracts for fee-based services to the total number of subscribers was over 100%.

The figures work out this way: 48% of i-mode subscribers had signed up for at least one fee-based service, and on an average, that 48% had signed up for 2.2 services each. Thus, when we had about 10 million i-mode subscribers, the cumulative number of customers for fee-based services was about 12 million (Figure 1.14).

The average fee for the services to which our users subscribe is rising as well. In July 1999, because the most popular fee-based content was Bandai's Kyarappa service, the cost was relatively low, averaging about ¥150. With 300 000 people each paying ¥150 monthly, the market for fee-based i-mode services was about ¥475 million per month.

Then more services began charging the maximum fee, ¥300 a month, for their content. Today, the average is nearly ¥200 per month. Thus, as of August 2000, the market for fee-based services had grown to ¥2.4 billion

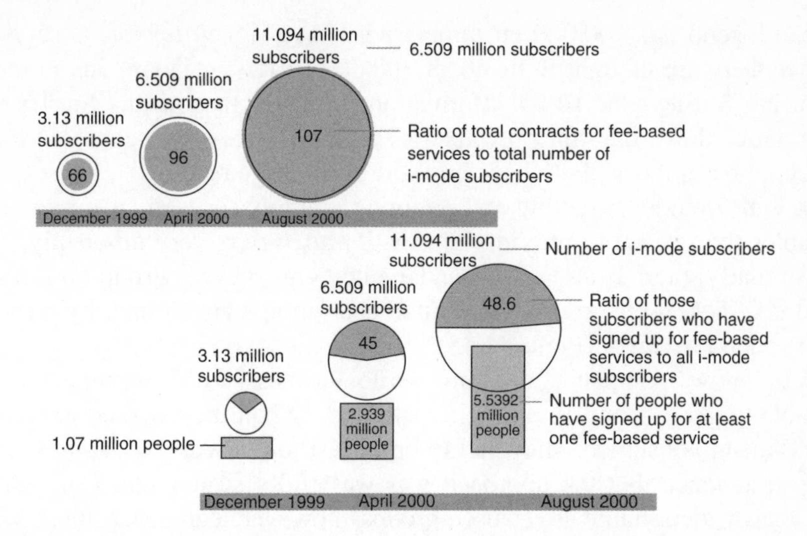

Figure 1.14 Astonishingly high ratio of subscribers to fee-based services. *Source*: DoCoMo.

a month, given that 12 million contracts for fee-based services at an average of ¥200 per month were in effect. That works out to nearly ¥30 billion annually – a huge market for fee-based services on a scale the wired Internet has not begun to match. And our market, we are sure, will continue to grow.

A major boost in the number of contracts for fee-based services, and in their average price, came with the launch of services for downloading ringtones. In September 2000, more than 6 million people had paid to be able to download their choice of melodies to announce an incoming call. Competition between service providers grew fierce, with over 20 on the menu offering ringtone download services. They then sought to differentiate themselves by offering new genres, and that in turn attracted more people to sign up.

1.6.2 Customer Retention Programs in Use

Another advantage of i-mode, from the point of view of service providers, is that they can lock in loyal customers. Banks that provide banking services, free of charge, have their eyes on that aspect of i-mode: making services available on i-mode is a new approach to customer retention. The Surflegend information site for surfers mentioned earlier is a classic example.

Surflegend has 100,000 customers who have signed up for its service. Since there are thought to be about 300,000 surfers in Japan, one in three is using Surflegend. To an information provider, that means Surflegend has nailed down one-third of Japan's surfers. Until i-mode, it would have been all but impossible to have any sort of message reach 100,000 surfers, but, with i-mode, attracting and retaining a highly focused customer base enables the service to provide information to surfers very efficiently.

Similarly, there is already a Bandai entertainment user group on i-mode and a karaoke user group, Xing. Such user groups are formed by content or service providers (Figure 1.15).

Until now, companies wishing to do business with consumers were unable to get a firm grip on their customers. When they wished to send a message to consumers, they had to broadcast to the world at large, knowing in advance that the approach was wastefully scattershot. Companies that have identifiable loyal user groups, however, can reach them with special offers and other devices. This access to defined user groups, more than income earned directly via i-mode, is perceived by many corporations as the real advantage of i-mode.

A third advantage is the opportunity for advertising and public relations activities. The recipe sites offered by Ajinomoto and Osaka Gas are good examples, and so are the restaurant sites offered by two companies in the beer and other beverages business, Suntory and Sapporo. To these companies, the act of developing their sites is a public relations activity that presents the corporate stance of 'creating a food culture'. This, of course, is a benefit that the wired Internet also offers.

The final advantage of i-mode is the ability to provide mixed media services to customers through links with the wired Internet. Net-based securities companies, banks, and other companies in the e-commerce field combine the capabilities of mobile phones and conventional computers quite effectively.

1.6.3 Stimulating Replacement Demand for Mobile Phones

The rise of i-mode also had a huge impact on the manufacturers of mobile phones because i-mode has generated replacement demand as subscribers traded for new types of phones. Mobile communications services and the hardware used (the mobile phones) are tightly bound together. When new services are made available, new mobile phones capable of utilizing the new functions are needed.

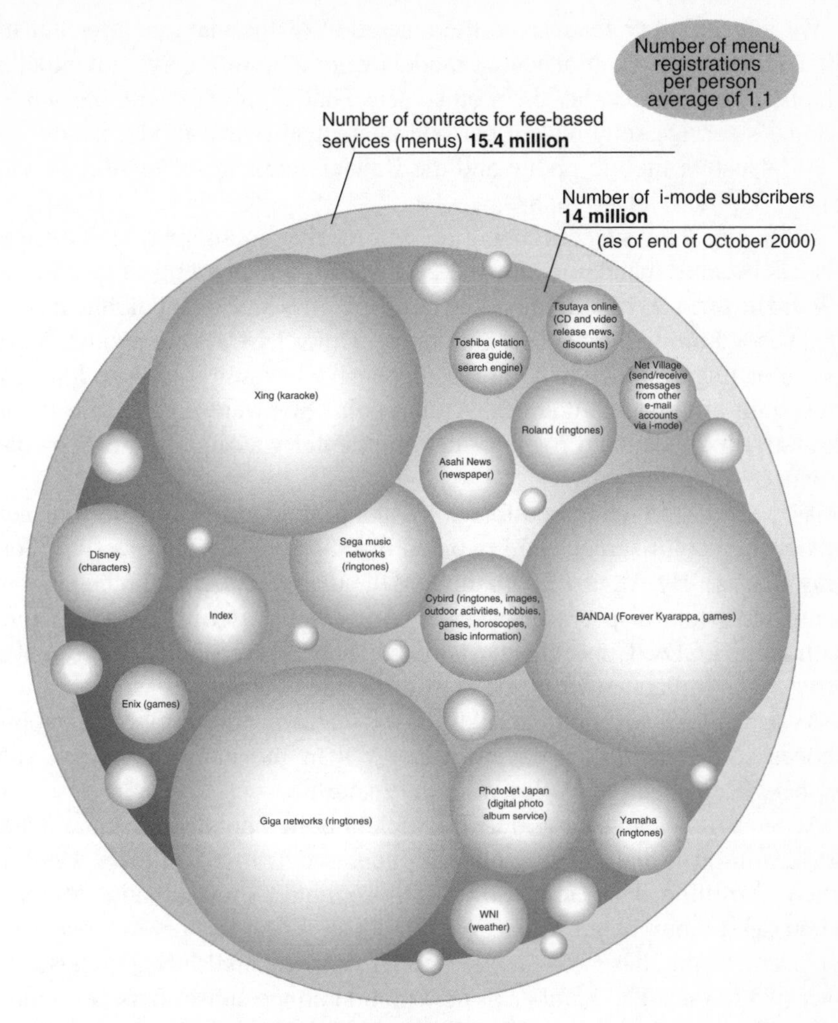

Figure 1.15 User groups on i-mode. *Source*: DoCoMo.

When we launched i-mode, we developed a separate product series for it, the 501i, apart from the basic 20x series of cellular mobile phone. In order to use i-mode, subscribers had to buy a 501i-series phone. The biggest difference between 20x and 501i was that we had added browser software to enable the 501i-series handsets to display Internet content, plus packet communications functions to enable data transmission.

We left the other features to the discretion of the manufacturer. Of the 501i series, only the Matsushita models were able to receive downloaded graphics files and display them on screen. That, combined with the start of Bandai's service sending out cartoon characters over i-mode, meant that the Matsushita mobile phone and the Bandai character-of-the-day service were both huge hit products.

Following their success, being able to receive and display downloaded visuals became mandatory for the 502i series, which went on sale in late 1999. The services for downloading new ringtones meant that mobile phones also needed multichord sound generators. Color LCDs were another story. Downloading graphics in color was expensive because it took so long and racked up high packet charges. At that point, we were unable to estimate how strong the demand for information in color was and did not see color as a necessity.

As a result, the manufacturers were again divided: two of the four new models had color screens and two had black-and-white screens (with four gray shades). By August 2000, though, the manufacturers who began with black-and-white displays in the 502i series were offering i-mode phones with color LCDs. Consequently, color displays were standard in the 503i series, which planned from the start to be Java-capable as well.

As a result of the gradual addition of features, the number of mobile phones sold became significantly larger than the number of new subscribers. Replacement cycles were very short.

According to a study by Merrill Lynch, mobile phone shipments totaled 20.42 million in fiscal 1997, but had increased 1.6-fold in fiscal 1999, to reach 33 million. Their breakdown of those figures shows that in 1997 new purchases outnumbered replacements, but in 1999, the reverse was true: replacement purchases of mobile phones outnumbered new purchases by over two to one. The number of new mobile phone subscribers per annum peaked in fiscal 1997 at 10.67 million and thereafter declined slightly. But replacement demand kept new phone sales strong.

1.6.3.1 Lightness Ceased to be the Decisive Feature

And here is something interesting: since we launched i-mode, the criteria determining which phones sell best have changed (Figure 1.16).

Until the start of i-mode service, the lightest mobile phones were the bestsellers. As people purchased their first mobiles, choosing the lightest phone simplified decision-making. Thus, all the manufacturers strove to

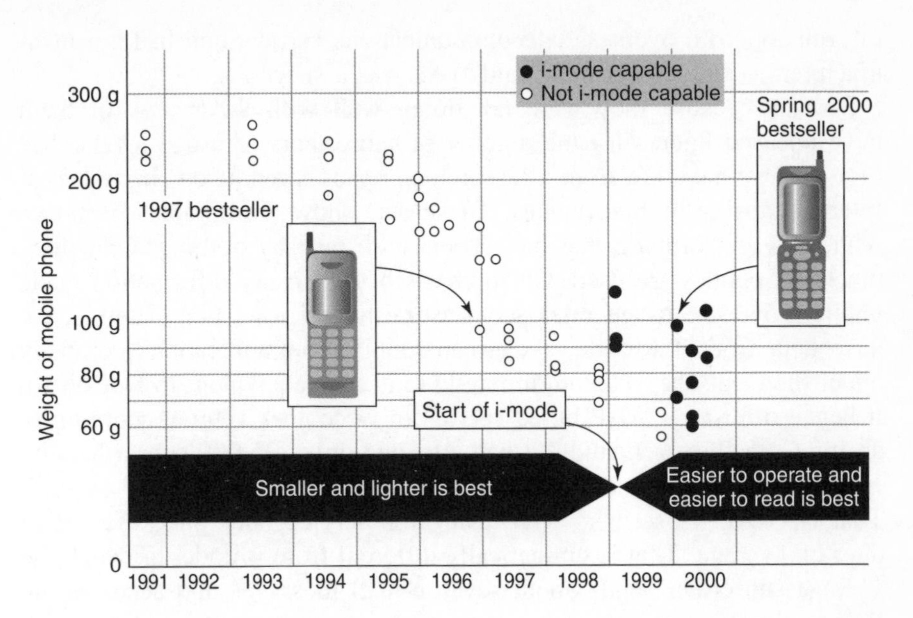

Figure 1.16 A change in which models sell best. *Source*: 'Game Renaissance', *Nikkei Electronics*, September 25, 2000, p. 174.

develop more compact, lighter models and managed to produce ultralight models weighing only 60 g.

But the launch of i-mode transformed preferences. In 1999 and 2000, the lightest models no longer sold best. The heaviest model in the 502i series, the NEC N502it, which folds, became a bestseller, for example. That would have been unthinkable earlier – but subscribers' criteria for choosing mobile phones had shifted from lightness and compactness to ease of use and content visibility.

1.6.4 Uniquely Successful: Service Providers in Other Countries Not Doing Well

In Japan, i-mode won widespread acceptance in a short period of time, but what was happening in other countries? Services based on the Wireless Application Protocol (WAP) proposed in Europe and the United States entered the market at about the same time as i-mode. All the overseas mobile communications business shows featured demonstrations of WAP. The market response to WAP, however, seemed extremely poor. Whenever I went overseas to negotiate partnerships for DoCoMo, I would exchange information with overseas telecommunications carriers and handset manufacturers – and almost none said WAP was a success.

Perhaps because they were not doing well with WAP, few of them have released figures for the number of subscribers or usage levels. The best data we have are from a research company's report on shipments of Internet-capable mobile phones. Those data show the number of phones with browser software that have been sold, but say nothing about how much the phones are used. On the basis of the scanty information made public, however, usage rates seem astonishingly low. For example, we have data from T-Mobile, a German mobile communications company under the Deutsche Telekom umbrella (and thus equivalent to DoCoMo). It began offering a WAP-based service in December 1999. According to an InfoCom Research study, it had attracted only 175,000 subscribers by June 2000, of whom only 35,000 a day were using the WAP service. That is, WAP subscribers were using the service only once every five days on average. That is dramatically different from i-mode, in which the average subscriber sends about seven e-mail messages and accesses the Web twelve times a day.

BT Cellnet, the largest mobile communications carrier in Britain, introduced a similar WAP service; as of September 2000, it had 400,000 subscribers. Of them, most use what BT Cellnet calls SMS (for Short Message Service). Very few indeed are true WAP users. Data from studies of WAP usage in the United States are, it seems, not yet ready for publication, but Merrill Lynch has predicted 20,000 mobile Internet subscribers in North America in 2000.

1.7 All Eyes on DoCoMo

Do we conclude, then, that people are not receptive to services like i-mode in other countries? I am often asked that question, and my answer is always no. There is definitely interest in i-mode-like services in other countries. DoCoMo is not alone in being eager to create the conditions to generate a data communications market, and other carriers are extremely interested in how DoCoMo has succeeded in Japan.

In fact, there have been more special issues on DoCoMo and i-mode in the media overseas than in Japan, and reporting on i-mode has been very positive. *Business Week*, an American business magazine, for example, made i-mode its cover story (see the color plates). It reported that while Japan had not been swift to take to the wired Internet, i-mode, which had started service just a year earlier, had quickly found wide acceptance. The article concluded, '... DoCoMo has latched first and best on to the mobile Internet, a technology with far greater potential than the other portable-electronics markets Japan has conquered. Calculators and camcorders do not carry with them an entire set of complex, Internet-based services, complete with new business models and lush venture-capital funding. All these and more come with the mobile Internet. Thanks to DoCoMo, Japan is out in front of the great land grab.' ['Amazing DoCoMo', *Business Week* (international edition), January 17, 2000, available at http://www.businessweek.com/2000/00_03/b3664010.htm.]

DoCoMo is constantly invited to attend mobile communications events. We seem to have become the poster child for the mobile Internet. I myself have given 20 speeches overseas in the past year. Add the speeches given by DoCoMo chairman Koji Ohboshi, president Keiji Tachikawa, and others, and the total given by DoCoMo must be well over one hundred. And when we introduce i-mode in such settings, the response is always huge.

At Wireless2000, a mobile communications conference held in the United States in February 2000, for example, none of the other keynote speakers aroused as much response as did our Keiji Tachikawa. Bill Gates of Microsoft was also a keynote speaker, but Tachikawa's speech was the bigger draw. During the question-and-answer segment of his presentation, he explained the i-mode fee structure, services available, and subscriber trends. The audience responded with applause and cheers to his presenting data on how i-mode is performing. They were particularly enthusiastic over our having attracted 4.5 million subscribers in just one year and over our fee structure, with subscribers paying packet communications charges and fees to access content on top of their basic subscription fees.

Not too long ago, telecommunications carriers and manufacturers in the West would look at tiny Japanese mobile phones and tell us, 'These won't go over well here because they don't suit our big hands.' But what is really going on? Compact mobile phones from Western as well as Japanese manufacturers are doing well in the US and European markets.

Consumer appetites do not vary that much from country to country – look at the Pokemon boom, which was global, or the Hello Kitty boom in Asia. Content is borderless.

Disney is another example: it launched its i-mode fee-based service in August 2000, and in just three months attracted about 500 000 paying customers.

Will i-mode spread overseas then? I am confident that it will, and soon.

Chapter 2
Concepts

2.1 Why Has Our Success in the IT Business Been so Overwhelming?

Because i-mode attracted over 10 million subscribers in just a year and a half, I frequently find myself asked to speak about how its unprecedented success – both in mobile communications and Information Technology (IT) – was achieved. At these presentations, someone always asks, 'What's the secret of i-mode's success?' I always bring up two points in reply. The first is that we applied an Internet way of thinking, not a telecom way of thinking, to everything from selection of technologies to proposing a business plan and acting on it. The other is that we understood the essence of the IT revolution and constructed and acted on a business model that followed the new, postrevolution rules.

2.1.1 It is no Longer a Telecom Age

What do I mean by an Internet way of thinking or a telecom way of thinking? Let us look first at the telecom way. Telephone service is a classic example. It works this way: a telecommunications provider provides the infrastructure (the network of phone circuits) and the customer uses it. It is a very simple model (Figure 2.1).

When a telecommunications provider wants to provide a phone service, the provider lays the infrastructure and provides a single service (telephone) that it designs from start to finish. It sets its own standards,

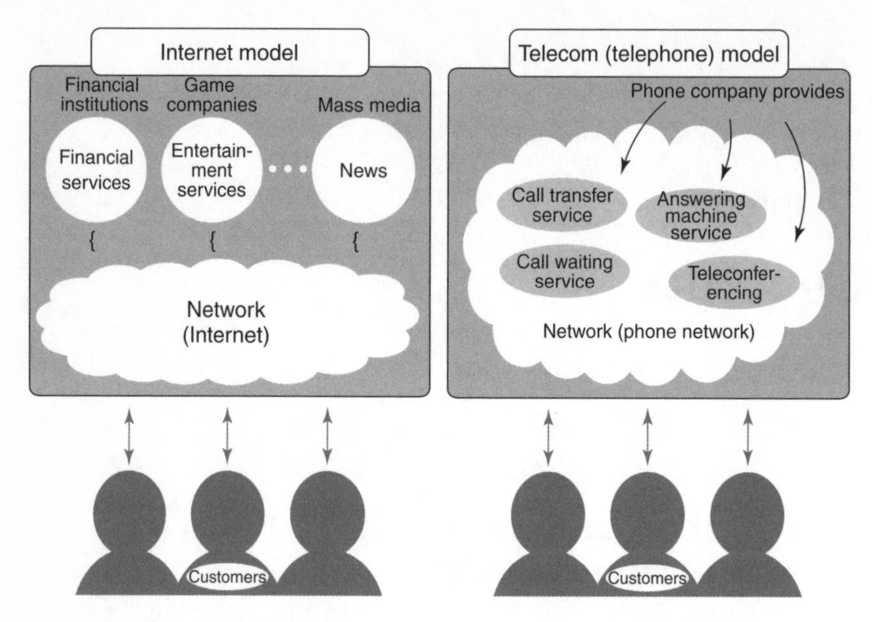

Figure 2.1 How the telecom and the Internet models differ.

including the specifications for the terminals (handsets) to be connected to its network, and it alone receives all the income (telephone fees) that providing the service generates. Because there are only two types of players – the provider and the end users, the subscribers – the provider is unlikely to think about using other already developed technologies or about ways to share income with others.

That model worked well when telephone service was the basic form of telecommunications. But it fails when a third player, a service provider, enters the picture, as in the Internet world today. The Internet model requires third parties to provide a variety of services, in addition to the provider who supplies the infrastructure and the end users who use it. Without someone out there thinking of attractive services in a business-like way, this model cannot fly.

The Internet model also assumes that the access point for the Internet need not be a personal computer: it can be a telephone, a television, or any number of other terminals. Thus, there can be a variety of ways to use the Internet and a variety of services available on it. It also means that the telecommunications provider's convenience cannot be the sole factor in defining terminal specifications.

2.2 Differences in Platform are Meaningless

Essential aspects of the Internet include its seamless connections to the end user and the device independence. The Internet does not assume a specific type of terminal – a telephone, for instance. It is a network built on a multiplatform premise (Figure 2.2).

That is, if you have a server connected to the Internet, you can use the shared platform known as the Internet to transmit content to personal computers running various operating systems, to personal digital assistants (PDAs), and to televisions and game machines such as the PlayStation – all can access it. The role of the telecommunications provider in the Internet age is not to decide what type of terminal is to be used but to make it easy to connect a variety of different types of terminals.

The seamless, device-independent Internet age concept means that technical differences between terminals are quickly ceasing to mean much. Before the Internet reached its current state, the war between the Macintosh operating system developed by Apple Computer and Microsoft's Windows operating system was clearly being won by Windows. But now that the Internet itself has become the killer application, the rules have

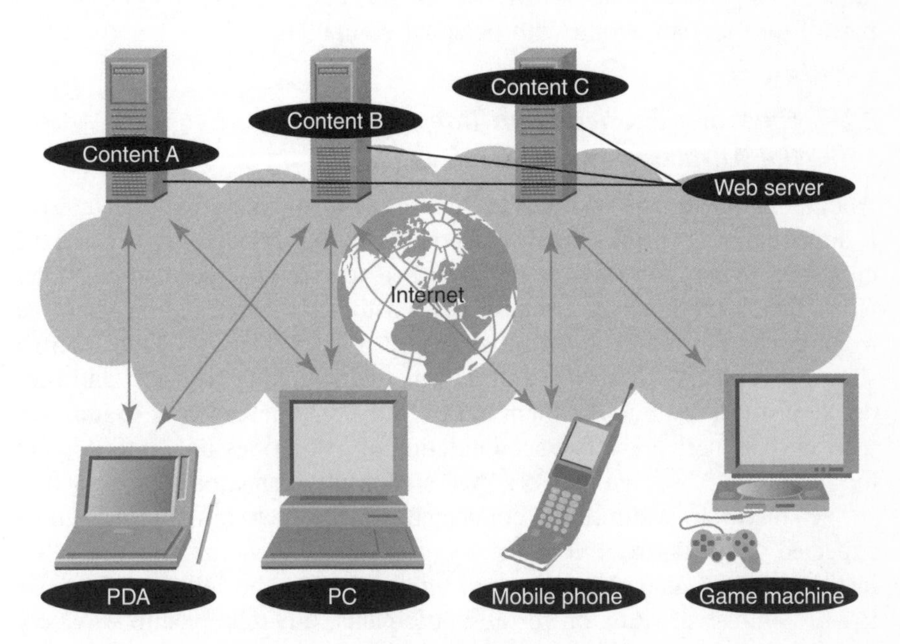

Figure 2.2 Seamless connection to any device.

changed so that it no longer matters whether the user is on a Macintosh or Windows system.

As we can see from the resurgence of Apple with its launch of the iMac, in today's world, competition from Windows is not the issue. What does matter is being able to use the Internet easily, whenever you want. It was here that i-mode – a latecomer among Internet terminals – found its point of entry to customer acceptance.

Who, you may ask, is that i-mode subscriber sending an e-mail message to in the middle of the night? The answer may be another i-mode subscriber – but often it is someone who accesses the Internet from a personal computer. In Japan, NTT's meter keeps ticking even for local phone calls, but it does offer a flat rate service that kicks in after 11 p.m. Consequently, more people go online from their personal computers after 11 p.m., and i-mode users start sending them e-mail messages then. Since, at that time of night, both can stay continuously online without the fear of huge telephone bills, they can exchange messages almost instantly, as if they were in a chat room or had an instant messaging service. After 11 p.m. is thus the time to have fun conversing by e-mail, using a computer or mobile phone. That is the virtue of a service that is seamlessly device-independent. That is why i-mode has naturally become an integral part of the Internet, along with personal computers.

2.2.1 Customer Participation Boosts the Attractiveness of Services

Another characteristic of Internet thinking is recognizing that the Internet is a medium that shrinks the distance between the service provider and the customer. Where to draw the line between service provider and customer is, in fact, very unclear, because one customer's participation can lead to increased convenience for other customers. The history of the Internet shows this clearly. It started out as a network linking research and academic institutions. Each of them linked to the external network to improve their internal networks. The accumulation of institutions linked to it led to the formation of a huge network that encompassed an enormous database.

The research institutions connected to the network because they expected benefits from it – just as it provided benefits for the other institutions that also joined it, and, ultimately, for the industry. Because it was a win–win situation for all participants, this relationship was very Internet-like.

Let us look at an example of an i-mode service that makes effective use of that characteristic – online games. Such games often list nationwide rankings, so that participants are able to compete to see who can score the most points. They tend to get quite obsessive about it. Each time a game participant's ranking goes up or down, it brings a smile or a tear – and yet more passionate involvement in the game. The customer is oblivious to the process, but his/her participation in itself makes the game more enthralling for others (Figure 2.3).

2.2.1.1 Customers and Service Providers are in the Same Team – That is Internet Thinking

Consider, too, that people in the online game industry tell me that their customers are quite uninhibited about sending them e-mail messages. One of the differences between an online game and a game distributed on CD-ROM, for example, is that the game designers can modify the story line in response to e-mails from the people playing the game. Players send in their requests to make the game more fun for themselves – and the game gets more interesting for everyone. Here, too, is a very Internet-like win–win situation.

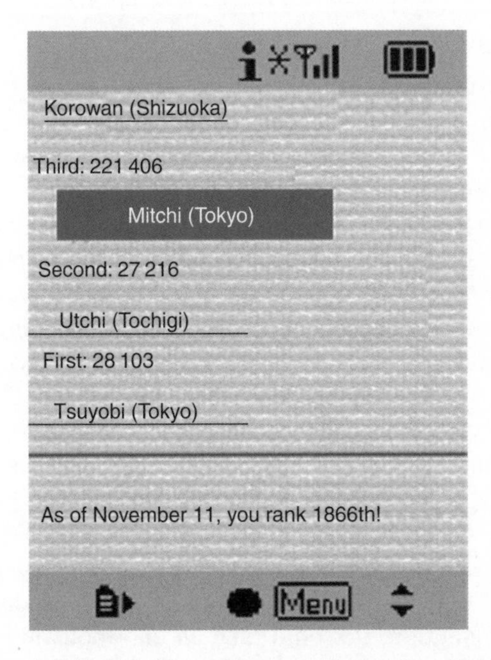

Figure 2.3 Nationwide rankings on 'Crazy about Fishing'.

We have seen something similar in the development process for Linux, the free operating system that is stirring up interest because it is perceived as posing a threat to Windows. Computer experts around the world had a hand in developing Linux, via the Internet – adding features, improving features, and bringing it to new levels of performance. That sense of working together is very Internet-like.

As these examples suggest, in the Internet age, service providers and their customers are engaged in a cooperative effort and, in doing so, form a harmonious whole. Or perhaps we could describe it as the age of complex systems (which I will describe in more detail later). In short, the day when one company, one technology, one industry provides all platforms is over. In the age of the Internet, launching a new business requires engaging in a cooperative enterprise that builds win–win relationships for all participants.

That is why, when I am speaking about i-mode and am asked for the secret of our success, I invariably talk about moving away from a telecom way of thinking to an Internet way of thinking. The mobile Internet is often described as a fusion of mobile communications and the Internet; this is true, and the shortcut to success was switching over to Internet thinking.

2.2.2 Alliances Create New Markets

Besides Internet thinking, I also point to what I see as the true nature of the IT revolution. The IT revolution has put an end to an age in which one major company, working alone, could carve out a new market. I argue that for two reasons.

First, thanks to the IT revolution, it is technically easy to organize cooperative efforts involving a number of companies. In this post-IT-revolution world, the basic rule is to use the *de facto* standard technology. Before the IT revolution, a company with its own unique and superior technology could pioneer a new market and emerge the winner. But today, when using the *de facto* standard is the name of the game, there is no way that a unique, proprietary technology is going to be accepted by potential customers. No matter how much a company asserts its technology's superiority, a technology that no one else supports is not going to win acceptance.

Second, it is now possible to build an ideal system, an ideal business model on a very tight budget – thanks to *de facto* standard technologies. Before the IT revolution, it would take an enormous investment in computer systems to act on an ideal business model. It was not unusual to have a great idea but be unable to bring it off because of lack of funds. Then came the spread of *de facto* standard technologies, thanks to the IT revolution the Internet brought about. It became possible to achieve whatever you wanted by way of a system. One example is intranets – network systems internal to a company built using the standard Internet technologies. If someone thinks, 'I would really like to be able to sign this electronically and then circulate it,' he can do so, easily and at very little cost, on an intranet.

In short, in our brave new world you no longer have to be one of the top companies awash in funds to have a chance to develop an advanced service.

2.2.3 Opportunities for Existing Businesses

Those who are racking up the greatest value by means of the IT revolution, however, are major corporations who were already doing business before the revolution arrived. Why? Using IT, they can make the businesses they have already developed even more efficient. If a major company really puts its efforts into making its existing businesses more efficient, it will leave no niches for start-up firms to fill.

One example can be found in the services that transmit cartoon characters for use as the default screen, or wallpaper, on mobile phones. These are an enormous hit, and they are provided by the biggest names in the cartoon character field, such as Bandai and Disney (see the color plates). Prior to the IT revolution, there were services distributing popular cartoon characters in the form of cards. But now a new means of distribution is available – the Internet and mobile phones – that makes it possible to send cartoon character images to customers without incurring either warehousing or physical distribution costs.

The cartoon character business has not changed, but by making it more efficient, these companies have created new value. Being able to transmit to mobile phones via the Internet means more of an opportunity for established businesses to become more efficient and expand, however, than an opportunity for start-up companies to launch new services. A new company could set up a system to transmit image files of cartoon characters, for example, but if it lacks an image library, it must go through the daunting process of acquiring usage rights to popular cartoon characters.

Thus, it is businesses already up and running that are best placed to make effective use of the IT revolution. Yes, the old economy has the pole position in this new race, but having the pole position does not mean you will win if your starting speed is slow or your engine falters or your driver is not skilful enough. Fall down in any one of those respects and the little start-ups will stop nipping at your heels and zoom on past you.

Still, I regard 2000 as a watershed year; from now on, the chances of big wins by start-up companies will decline. The major corporations from the old economy will take full advantage of the benefits of the IT revolution. I am not saying that start-up companies have no chance of success at all, but I think they would be well advised to work toward having a big company recognize the value of their technologies and then to cooperate with it in making its existing businesses more efficient. Start-up companies have a tendency to want to earn names for themselves for stealing a march on the big players or grow to rank with the major corporations. But if they forget about glory and focus on practical achievements, there are heaps of opportunities out there.

2.3 Why is the Win so Overwhelming?

Internet thinking and understanding the true nature of the IT revolution were two key points we kept in mind while developing i-mode. In fact, we devised all our strategies in terms of what is known, in the academic world, as complex systems theory. We realized that we live in an age of complex systems, with the Internet linking a multiplicity of elements together. That perception led us to the idea that in an IT business such as i-mode, we could fully employ complex systems theory. Where, though, does complex systems theory come from?

At the start of Chapter 1, I said, 'IT businesses grow far more than expected or do not grow at all.' As we launched i-mode, which is indeed a new IT business, I wondered why, in fact, things were so black or white for new IT businesses.

Cast your mind back to other IT businesses and you will realize that companies with technical superiority have not always won user support. The Video Home System (VHS) versus Beta video format war is one illuminating example. As the result of a ferocious battle between Sony, with its Betamax format, and the proponents of the VHS format, Matsushita Electric Industrial (manufacturer of Panasonic and other consumer brands) and Victor Company of Japan (JVC, another consumer electronics giant), VHS won a victory that eliminated Beta from the market. A similar competition developed between Apple's Macintosh and Microsoft's Windows operating systems for personal computers. The result was a crushing victory for Windows, which now dominates the market.

In both cases it was not possible to say that one side or the other was absolutely superior in terms of technology. In fact, the technologies that ended up a minority in the marketplace tended to have the support of engineers and other specialists. Nonetheless, the result was an overwhelming victory for one format, one operating system. Why? If technology was not the deciding factor, what was? And why were the victories so one-sided?

2.4 Life Today: Complex Systems

So many phenomena are at first glance inexplicable, impossible to understand – and not just in the IT business. The many phenomena that do not seem susceptible of a logical explanation include stock market booms or busts, the collapse of nation states, or corporations' prosperity or decline.

At first glance, those examples may seem like apples and oranges, but they do have something in common: they are all events that occurred in complex systems. There is a field of scholarship, complex systems theory, that attempts to find logical explanations for phenomena that occur in complex systems. In many cases, applying that theory can explain events that seemed inexplicable.

In a complex systems theory model, all phenomena are composed of many constituent elements of many different types, all interacting with each other. What is interesting is that those constituent elements can, in response to a stimulus, exhibit self-organizing activity as a collective whole, despite no conscious awareness on the part of the elements. Self-organization means that the system comes to possess a certain order or directionality. The stimulus for self-organizing activity can come either from within, from the elements that make up the system, or from outside.

2.4.1 Self-Organization in Geese

A model often used to explain self-organizing activity in complex systems is a flock of geese in flight. We observe them taking off at sunset. At first, the geese fly in no particular order, but gradually they organize themselves into a graceful inverted V shape, with one goose flying at its tip. The formation may change shape somewhat over time but it will not fall apart. While the formation is clearly organized, the individual geese flying in it have no concept of the formation as a whole and the goose in the point position has no conscious desire to lead the flock nor is it sending orders to the flock as a whole. In fact, what each goose is thinking of is not the flock or the formation but its relationship to the neighboring geese and to the earth's surface.

Using complex systems theory to run a computer simulation of a flock of geese in flight, researchers found that they could reproduce the V formation by inputting only three commands for each individual goose: 'fly in this direction', 'maintain this angle and separation with respect to your neighbors when in flight', and 'maintain this height above the earth's surface'. If you input precisely those commands in a number of virtual geese and program them to take off together, after an initial period of chaos, they will be flying in a stable, elegant formation. And if your program includes obstacles down on the ground, the formation as a whole will respond to them as though it were a living thing, but will not disintegrate (Figure 2.4).

This commonplace phenomenon could not be explained by conventional analytic approaches. Conventional thinking could not explain how to maintain leadership over the whole flock without programming the flock as a whole. But with complex systems theory, it is possible to account for the flock's acting in a purposeful, coherent way despite inputting

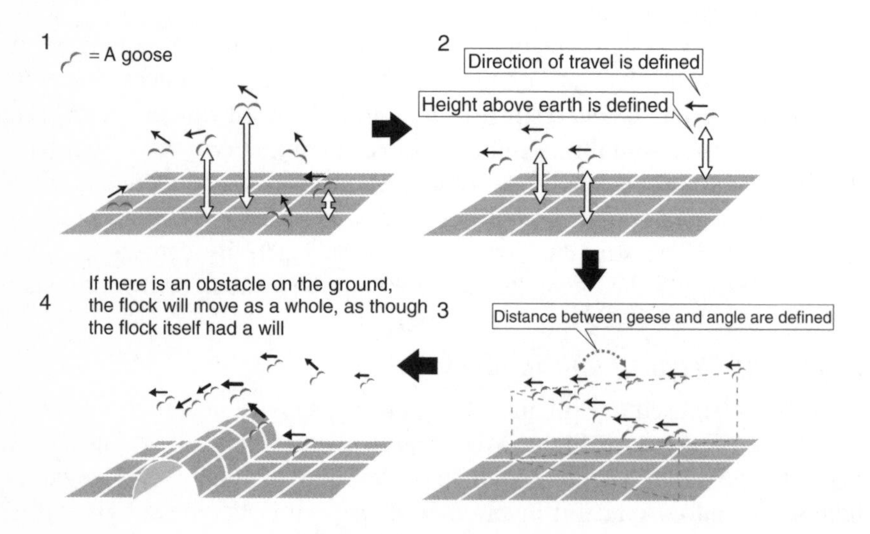

Figure 2.4 Geese in flight.

information only for individual members and not providing a program for the flock as a whole.

Much of human economic activity can be explained in ways similar to the geese model. Economic activity in human society consists of individual elements working toward optimizing their situations relatively: looking at what other companies are doing, for example, and deciding what one's own company will do. That touches off organized behavior, organized economic activity, and, by extension, economic cycles, within a nation state or globally – without anyone attempting to control the system as a whole.

2.4.2 Evaluating a Service in Terms of the Service as a Whole

How can we use complex systems theory to explain the one-sided victories that we observed earlier, in the competition over video cassette recorder (VCR) formats or personal computer operating systems? In both cases, if you evaluate the separate technologies that make up each competing standard, there is no basis for telling which one would win. But in complex systems theory, we include not just technology itself but the peripheral elements – the service as a whole – in explaining the outcome.

Consumers, when comparing services using different formats, do not compare the separate technologies that are elements of each; they look at the service as a whole and evaluate it. That is, when choosing a VCR format, the customer is not going to compare the various video specifications. What the customer looks at is the advantages that can be received from choosing one format – whether a wide range of videos are available recorded in that format, which format friends, with whom they might share videos, have, and so on. Those benefits become the key points in making a choice. When the video rental shops stock only VHS tapes, the VHS format would have achieved a complete victory.

A similar process applies to personal computer operating systems. While the market for personal computers was small in scale, people argued over the features of the operating systems themselves. But when the market grew rapidly larger and personal computers became mass market items, what decided the choice was how complete a line of operating software and peripherals was available. It would be an overstatement to assert that most personal computer users are not interested in which operating system is technically superior. Rather, what they do care about in making choices is which operating system the applications they want to use run on.

In both cases, applications flock to the platform that wins, and the more applications it attracts, the greater will be the convenience to users, and more users will choose that platform. That virtuous cycle produces an overwhelming victory for one side in the competition.

The repeated action of that virtuous cycle is why there are no indecisive victories in the IT industry. Those overwhelming victories are the result of what is called, in complexity theory, diminishing returns.

2.4.3 One Technology Cannot Lead a New Service

In looking at an IT business from the point of view of a provider, the key to success in trying to start a new business is finding ways to motivate other technologies, other companies, and other industries to participate. One technology, one company, or even one industry alone cannot provide the lever to move the world.

Technologies, economies, and communities are becoming increasingly borderless. The spread of the Internet has accelerated that trend. Greater borderlessness generates greater social complexity and more factors affecting the market. It is quite hopeless for one corporation to try to drive the whole market on its own. It is because of this context that I thought that complex systems theory could be effectively applied in the age of the Internet.

What must one do then to launch a new service and create a huge trend that will sweep the whole market along with you? The necessary condition for that success is building a service model that makes it simple for as many technologies, corporations, and industries as possible to choose voluntarily to participate in the service.

In terms of the video format example, the participants included producers of videos, video deck manufacturers, and video rental shops. And, of course, those who view videos – consumers – should be counted as participants as well. Building a service model that consumers will think has advantages for them is the driving force that creates market trends.

The basic theory of complex systems is, thus, simple. When making a choice, we act in terms of our assumptions about what impact that choice will have, and where. But those assumptions are not reliable predictions. In fact, the complex systems approach would be to assume that the outcomes will not be as predicted. In this complex world, we must be ready to observe outcomes contrary to our predictions and to make subsequent decisions flexibly.

2.4.4 DoCoMo's Role is to Coordinate the System as a Whole

Inducing all of a varied array of indeterminate elements to move in the direction in which one wants to go: that, as the model of the flock of geese makes clear, is indeed a very complex systems way of thinking.

In our i-mode service, the component elements are a long list of corporations, including service providers and manufacturers. To get the service running on track, it was necessary to build a service model that they would all think advantageous to themselves. Happily, DoCoMo was in a position to coordinate the service as a whole – mobile phones, networks, systems, and content (Figure 2.5).

Complex systems theory uses the concepts of positive feedback (virtuous cycles), increasing returns, emergence, and self-organization.

Positive feedback: An increase in subscribers promotes more service providers to participate, which attracts more subscribers and thus generates new positive development.

Increasing returns: As a result of positive feedback, one side in a competitive relationship will reap an overwhelming victory.

Emergence: The small actions of components generate new technologies and methodologies. In the Bandai example described above, Forever Kyarappa is a classic example of emergence. Emergence is evolution at the level of individual components.

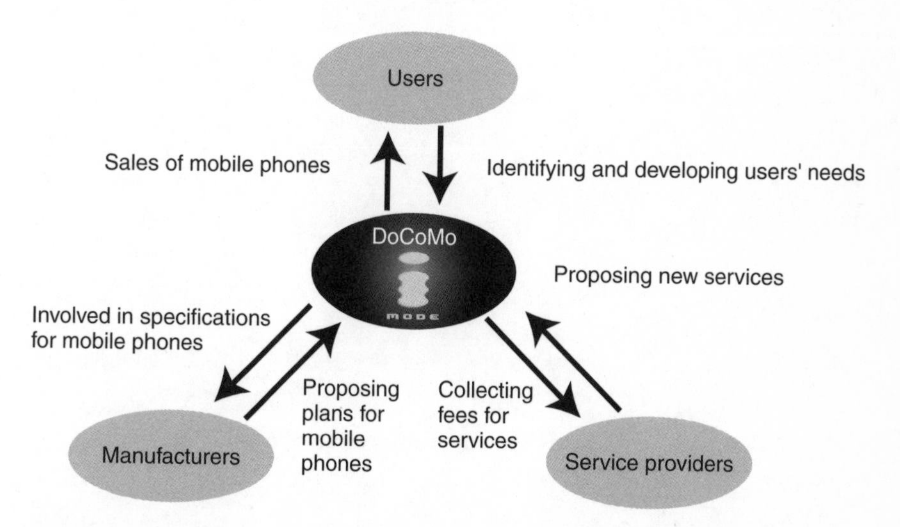

Figure 2.5 DoCoMo is the pivotal player for service providers and manufacturers.

Self-organization: Evolution of the organization as a whole. A good example of self-organization would be that over 320 banks now participate in i-mode to provide mobile banking services.

In designing i-mode, we applied these concepts to building the actual business model and technical platform. In the next chapter I will explain how we applied complex system theory concepts to i-mode services.

Chapter 3
Practice

3.1 Most People are Conservative; They Reject What Seems too New

I joined DoCoMo's i-mode project in September 1997. At that point, I raised three points for discussion with the other project team members (see Appendix 2).

The first of those points was where we were going: how far we would be trying to take i-mode. The second was a theory of the steps we should follow: what steps DoCoMo should take at each stage in i-mode's development. The third was a concrete strategy for the content portfolio and our relationship with content providers.

That position paper could be called the basic design for i-mode. The basic i-mode strategy has, since then, been unfailingly guided by the ideas I expressed in that paper.

3.1.1 There Will be a 'Wallet PC' Someday

The vision of where we were going, my first point for discussion, was, quite simply, the 'wallet PC' that Bill Gates wrote about in his book *The Road Ahead.* It would be a computer small enough to fit in the palm of your hand, yet could hold everything you need to take with you when you go out. Cash, credit cards, prepaid cards, address book, concert or airline tickets – all would be combined in one small computer. That was the concept.

The scenario Gates had sketched for delivering the wallet PC concept was to make existing computers more compact and add network functions to them. Gates wrote that notebook computers will become thinner and thinner, until eventually they are about as thick as a pad of stationery. Pocket-sized computers with color screens the size of a photograph will be as commonplace as wallets are today. The wallet PC will, he argued, incorporate the functions of mobile phones and pagers as well.

What then was the i-mode approach? Our goal was the same as Gates', but we were starting from a different point. We were going to add computing functions to a device (the mobile phone) that was already connected to a network, develop it, and get close to the functions of a personal computer. In fact, at that time, I thought, 'We may be able to beat Bill', because compared with shrinking a personal computer while adding network functions, beefing up a mobile phone with computer functions, our approach, would get us to our goal far faster.

3.1.1.1 Add-Ons Stimulated the Appetite to Develop

My reasoning was that shrinking a personal computer would involve lopping off features, and that approach would be counterproductive in stimulating technical progress. It would be painful to have to decide which of the functions you had worked so hard to incorporate into a computer should be dropped. But we were taking an add-on approach with our mobile phones, and that would facilitate technical progress. When your approach involves cutting features, there is always the feeling that you are retreating. When you are adding features, you are moving ahead. That, I thought, would stimulate the technical people's inventiveness and speed up the pace of development.

We also had the advantage of sheer numbers: the potential unit sales of mobile phones far outstripped those of personal computers. Technology tends to converge on products that can sell in large quantities. That means that manufacturers would choose to invest their human and financial resources in developing new mobile phones and components, products with unique features no other company could offer. Thus, for example, the shift to color screens happened faster with mobile phones than with personal computers, simply because manufacturers could count on volume sales. A large population makes it easier for a positive feedback loop to get started.

That advantage of mobile phones would, I thought, be particularly marked in Japan, where the mobile phone to personal computer ratio is especially lopsided. Personal computers are not seen as a necessary part of life in Japan to the extent they are in the United States. Perhaps because we think that handwriting reveals character, Japanese culture tends to value handwritten documents, even for business documents. In the United States, by contrast, the use of typewriters and computers is widespread. It has become almost necessary to use a personal computer to prepare your personal income tax return.

Japanese and American perceptions of the convenience of computer networks are also radically different. In the United States, it is now perfectly normal for junior high school students, assigned to write a report, to use personal computers and the Internet to search through online catalogs for reference materials. Library catalogs are only one example of the many types of information that have been digitized and made available on networks for use in all aspects of life. The introduction of the Internet was a natural, indeed almost inevitable, development.

In Japan, personal computers and Internet use are not yet as widespread. But everyone has a mobile phone. That gave us a vastly larger base of mobile phones.

3.1.1.2 One Industry Alone Could not do It

In thinking about the wallet PC, I also realized that no single industry, no matter how hard it tried, could bring it off. Enabling this mobile terminal to have payment functions, for example, would require securing the understanding and cooperation of banks, credit card companies, and other types of businesses as well. The feasibility of attracting other industries to participate in the project depends again on scale. When a service provider is making a decision on the timing of its entry, it looks first at the total population to whom the service would be provided and then at the percentage likely to subscribe, to estimate how many customers it is likely to attract. Obviously, the larger the total population, the better. Here, too, the huge numerical advantage of mobile phones over personal computers was a strong plus point.

We also believed that mobile phones had an advantage in terms of consumer acceptance. Many people who cannot use a computer do use mobile phones – but the reverse is not true. Assuming it was possible to do the same things with a mobile phone or a pocket-sized computer, we were confident that people would have less resistance to using the mobile phone.

Our vision for what mobile phones would become is also a perfect fit with the network computer concept advocated by Scott McNealy, CEO of Sun Microsystems. McNealy envisioned a system in which data on networked servers could be seamlessly accessed from all sorts of terminals. Sun had developed its platform-independent Java programming language for developing applications without worrying about differences in operating systems.

In addition, the Internet-based advertising business in which I had been working before I joined i-mode was also something I had thought I could pull together on conventional (non-i-mode) mobile phones. The concept was that the customer would be able to receive product information that suited his or her preferences and demographics, free of charge, through mobile phones and the Internet. The advertisers would be able to send product information to the people who really wanted it – that would be more efficient. And DoCoMo, positioned in between the customer and advertiser, would be able to function effectively as the gatekeeper mediating between them.

3.1.2 How to Kick-Start the Process?

The second point I raised concerned the steps to be taken. How could we, first of all, attract a million subscribers? What would it take to hit ten million subscribers? What then would our next step be?

The key question was the stimulus that would touch off this process. If we did not get the first million subscribers, then there would be no question of five million, much less ten. Here we were in the same position as any businessperson trying to start a new service and puzzling over how to build a mechanism to get the process rolling.

Our headache at i-mode was a familiar one: if we did not have any services to offer, we would not attract subscribers, and if we did not have subscribers, we would not be able to attract service providers. It was the classic chicken-and-egg problem. And if we picked the wrong stimulus with which to start up the process, we could easily end up with egg on our face: a negative feedback loop, not the positive one we needed. Then the i-mode market would never get off the ground.

Actually, the most important point enabling i-mode to win a million subscribers in a flash was our making sure that we already had an extensive, attractive lineup of services (content) available when i-mode started. The deciding factor was that, at a point when we had almost no subscribers, we did offer a rich array of extremely high-quality services. In fact, 67 of Japan's leading corporations had agreed to participate in i-mode as it was launched, when it had, of course, zero subscribers (Figure 3.1).

Category	Service	Sites	Companies
Mobile banking	Balance inquiries, transaction details, bank transfers, and information	21	Asahi Bank, Bank of Tokyo-Mitsubishi, Dai-ichi Kangyo Bank, Daiwa Bank, Fuji Bank, Fukuoka Bank, Fukuoka City Bank, Higo Bank, Hiroshima Bank, Iyo Bank, Kita-Nippon Bank, Kiyo Bank, Nishi-Nippon Bank, Ogaki Kyoritsu Bank, Sakura Bank, Sanwa Bank, Sapporo Bank, Shiga Bank, Sumitomo Bank, Suruga Bank, Tokai Bank
Mobile trading	Stock price information, market information, buy and sell orders	2	Daiwa Securities, Nikko Securities
Credit cards	Information on special offers, credit card bill information	4	JCB, Sumitomo Credit, DC Card, UC Card
Life insurance information	Information on procedures and processes	5	Sumitomo Life, Dai-ichi Life, Nippon Life, Meiji Life, Yasuda Life
Airline information	Inquiries about seats available; making reservations, inquiries about mileage totals	3	ANA, JAS, JAL
Hotel reservations	Inquiries about room availability, reservations	2	JTB, Pleco
Discount travel information	Retrieve information on inexpensive travel, make reservations	1	Open Door
Train connections	Information on where to make transfers and restaurant guides	2	JR East, Toshiba Ekimae Adventure Club
News and sports news	News in general, sports news, entertainment news	5	Asahi News, Jiji Press, Hokkaido News, Mainichi News, Yomiuri News
Share price information	Share price information	1	Japan Telemedia Service
Weather forecasts	Weather forecasts	1	Weather News
Ticket information	Retrieve concert information, make reservations	3	Ticket Saison, Pia, Lawson Ticket
Real estate rental information	Retrieve information on rental properties	1	Able
Recipes	Names of dishes, recipes	2	Ajinomoto, Osaka Gas
Karaoke	Retrieve names of karaoke songs, karaoke centers, new songs	1	Daiichikosho
FM station information	Retrieve names of songs, program information, hit chart information	2	FM802, J-Wave
Book sales	Retrieve information on books, purchase books, bestseller information	1	Kinokuniya Bookstore
Dictionaries	English-Japanese, Japanese-English, several Japanese dictionaries, thesaurus	1	Sanseido
Games	Online games	1	Bandai
Local information	Restaurant guides, movie theatre information	2	Yellow Pages, Bay Area
Fortune telling	Fortune telling	2	Animo, Index, Telesys Network
Phone numbers	Retrieve business phone numbers, by region	1	NTT
Other	Surfing informaion	2	FM Chuo, Cybird

Figure 3.1 The starting lineup of content providers (the 67 i-mode pioneers).

3.1.2.1 Language Selection: A Keystroke

What was it that freed us from our chicken-and-egg problem, our decisive move? It was our adoption of HyperText Markup Language (HTML) as the language for i-mode sites (Figure 3.2). That choice of language meant that we were using a *de facto* standard – and it was the first step in putting our i-mode strategy into effect.

HTML is the standard language for displaying information on the World Wide Web. There were other alternatives for the i-mode markup language, but we based our decision on the impact the language choice would have on other elements making up the i-mode system. As explained in the previous chapter, we used what I have called Internet thinking and complex systems theory as our basis for making decisions about i-mode.

That is, instead of considering only which language was superior, we ran simulations of how service providers would react depending on which we chose. The result, our choice of HTML, was the winning move that set the positive feedback process in motion.

We chose HTML for i-mode because it is the standard for marking up content for the Internet. Strictly speaking, the basis for the language

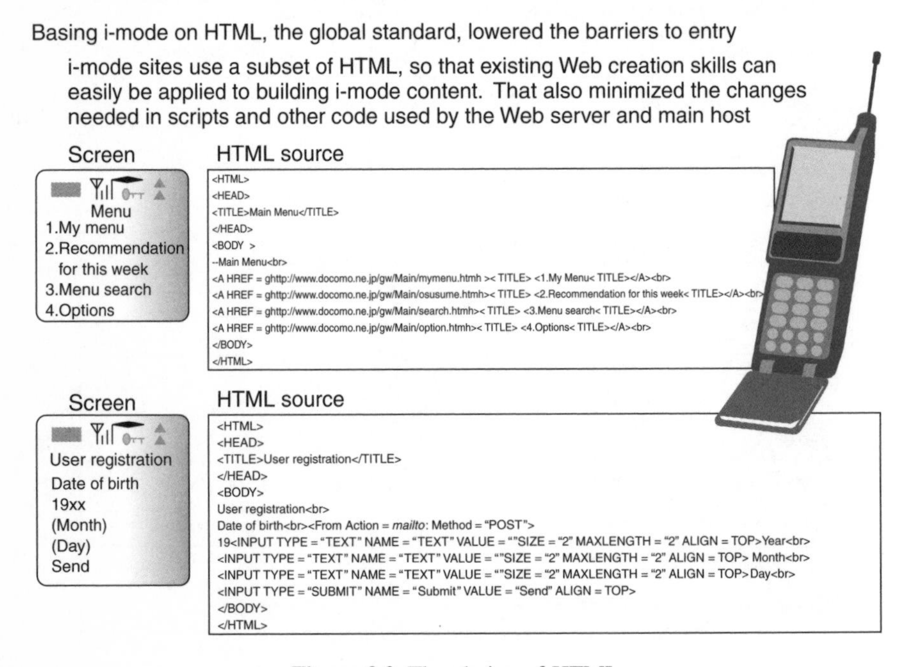

Figure 3.2 The choice of HTML.

used on i-mode is something called CompactHTML, a subset of HTML that had been proposed to the World Wide Web Consortium (W3C), the standards organization for Internet content. We expanded CompactHTML slightly to improve the user interface. But what we use is, all intents and purposes, basically HTML.

3.1.2.2 Our De Facto *Standard Technology Lures Content*

Our objective in using HTML was clear: we wanted as many service providers as possible to provide content on i-mode. Enormous amounts of content were already available on the Internet. If i-mode also used HTML, it would take those already providing content on the Internet only a little extra work to make the few modifications needed to send that content to i-mode subscribers.

What was important to bear in mind was that the population of people using HTML to produce Web content was already large. As anyone who has done so will agree, knowing a little HTML makes it possible to produce i-mode content with ease. Using HTML, then, dramatically lowered the barriers to producing i-mode content.

Not everyone agreed with that decision. Some argued that HTML is too inefficient for wireless transmissions or that technically superior choices existed.

In fact, at the time, Nokia, Ericsson, and Motorola, the top three manufacturers of mobile phones in Europe and the United States, were central in proposing an alternative to HTML, the Wireless Application Protocol (WAP). WAP is a data transmission protocol that is optimized for wireless transmission use, and telecommunications companies throughout the world were working for its adoption. DoCoMo is in fact a member of the WAP Forum, the body promoting the technology, and was well aware of what was happening with WAP.

When we looked into WAP, we learned it is indeed a very simple protocol that is specialized for the wireless environment, where transmission quality is not always good. WAP simplifies interaction between servers and terminals, minimizing the likelihood of transmission errors.

3.1.2.3 A Lesson from a US Study

While WAP had its virtues, we were concerned about the difference in markup language. While WAP is actually a collective name for a group of protocols, the markup language used with it is something called Wireless

Markup Language (WML). It is a language utterly different from HTML; they have nothing in common.

Using WML as the markup language for i-mode content was an option we considered. We had manufacturers supporting WAP and WML and who were proposing that we use it. But in the end, we did not.

Adopting WML would have made more work for our service providers, and, we thought, we would be unable to attract as much content as we wished. With WML, service providers would have had to invest twice – in their existing Internet sites and in the new i-mode format. In addition to HTML programmers to mark up the content and their existing Web servers using HTML, they would have had to line up servers and programmers for WML. Do you think they would have leaped at the chance?

In fact, while considering whether to use HTML or WML, I went to the United States with some representatives of the manufacturers. We had heard that AT&T Wireless, a US mobile communications provider, had begun offering PocketNet, an information service for mobile phone users, and we wanted to try it out for ourselves.

The AT&T Wireless service uses what it calls Handheld Device Markup Language (HDML), a language based on WML developed by a US start-up company. In any case, it too is utterly different from HTML as we know it on the Internet.

When we tried it out, we found that the only items on the PocketNet service menu were ways to check whether flights were delayed and to find telephone numbers of restaurants. The menu was so limited that I suspected no one would use it. Upon asking, I learned that they had only about 5000 subscribers, which was about what I had expected. I headed back to Japan doubly convinced that the depth and the breadth in the content we offer would be the key to success in popularizing our service.

When I returned from that fact-finding trip, we decided to use HTML for i-mode, and we started hustling to get content providers lined up.

3.1.2.4 Kindergarten English? or French?

Choosing HTML, the standard language of the Internet, was the right decision. When we presented the i-mode service concept to likely content providers, almost all of them saw the point when we explained that we were using HTML as the markup language. As a result, on February 22, 1999, when we launched i-mode service, 67 companies simultaneously launched their services on it. Since then, we have had a stampede of companies wanting to provide content on i-mode, and the wealth of content they are offering has won more subscribers. Most of those companies also provide content on the wired Internet.

The low barrier to entry set by our adoption of HTML applies to both sites on the official i-mode menu and the many other voluntary sites. Unlike the official sites, there are no links on the iMenu for voluntary sites, and DoCoMo does not collect fees for those who provide those services. Nonetheless, an extraordinary number of voluntary sites have been launched for i-mode. Several billing systems for the voluntary sites have also been developed, and more keep coming. All this has happened because we decided to use the same language as the Internet.

I have been using foreign languages as a metaphor for comparing i-mode HTML and WML. If HTML is English, I say, then i-mode HTML is English as spoken by kindergarten kids. WML may use the same alphabet, but it is French – you have to learn it from scratch. It is obvious which would be easier for an Internet content provider who had already mastered HTML.

Our choice of markup language was one based on Internet thinking. In a purely technical comparison, the advantage would go to WML. But whether the service using it would take off or not is a separate question that cannot be answered by mere comparisons of technology. The question is: Which would get a content developer moving, HTML or WML? Making that our criterion was a winning stroke in our success with i-mode.

3.1.2.5 Thousands of Ringtones in No Time

In addition to the markup language, there are several other component technologies for displaying content on i-mode. In each case, we followed the same line of thought as when we chose HTML. That is, our priority was to make the barrier for content providers as low as we could.

For example, we adopted the Graphical Interchange Format (GIF) for our graphic data format. GIF is a format for graphics already commonplace on the Internet. There were other graphics formats that, in terms of compression or licensing, were better options than GIF. The J-Phone Group had, for example, adopted the Portable Network Graphics (PNG) format for Web sites on its J-Sky service for just those reasons.

Ultimately, though, we looked at the statistical data on what is used most on the Internet as the basis for deciding to adopt the GIF format. Today a wealth of graphic content is available on i-mode, including the highly popular services transmitting the cartoon character of the day. That is because we used the GIF format that is widely used on the Internet. To service providers, there was no need to redo their graphic data; they could easily incorporate their existing graphics into their services. Because there was no need to worry about technical differences between i-mode and the personal-computer-based Internet, content providers could focus their creativity where we need it and our subscribers want it – on better services.

We applied the same thinking when it came to one of the distinctive features of mobile phones – the ringtone download function. We used Compact MIDI, a subset of the familiar Music Instruments Digital Interface (MIDI) music description language format.

MIDI has a considerable track record for use apart from the Internet, in karaoke-on-demand services. Companies like Daiichikosho, which provide karaoke-on-demand, already had huge libraries of music data in MIDI format. With just a little massaging of their data, those companies could use them in ringtone download services on i-mode.

Each of the karaoke-on-demand companies has music libraries with tens of thousands of tunes, so that our subscribers are certain to find tunes they like, whatever their favorite genre is (Figure 3.3). We at i-mode of course

Figure 3.3 Ringtone Melody Giga, from Giga Networks.

wanted to be able to include those assets in our service, and to the karaoke companies, being able to use their assets in ways other than karaoke was also attractive.

Suppose that DoCoMo had proposed its own format for music files. No matter how many subscribers i-mode attracted, the potential market on i-mode would not have covered the cost of converting huge volumes of music files. By adopting MIDI, we were able to establish what has become a huge market for ringtone downloads, in no time.

3.1.2.6 Unusual Phones do not Sell

When it came to technical decisions other than those related to content, we again gave greatest weight to the impact our decision would have on other components of the complex system we were constructing, just as we did in choosing HTML. Those technical elements, for i-mode, can be broadly divided into technologies for mobile phones and those for the network system. The external components to be taken into account are different for each category.

In developing an i-mode-capable mobile phone, our concern was that the result be something subscribers would accept readily. Apart from a few technology buffs, most people are quite conservative. They do not want to stand out from the crowd. That tendency is clear in mobile phone sales trends. Rather restrained, serious-looking mobile phones – silver ones, for example – sell well. Yes, the manufacturers also offer them in racy colors, just to have a varied line to display, and some people are attracted to them while looking at phones in store displays. But, while they may handle the more colorful phones, they rarely buy them. They buy the silver phones.

Similarly, no matter how much you advertise the excellence of its features, if a mobile phone is in a format that is very different from what is currently in use, people will not buy it. A mobile phone that is strikingly different in either color or format will not be a bestseller. (Nor, of course, is success with an un-cool-looking model possible.)

At the start of development, when we explained to manufacturers that the proposal was to build mobile phones with Internet-connection capabilities, they indicated that what they were imagining was ultra-high-tech phones or elite phones for executives. But most consumers would reject these phones if we stressed high performance and multiple functionalities. A few geeks would take to them but not the rest of the market, and that would be that – a flop for i-mode.

At that time, DoCoMo itself had marketed several models that combined the functions of a PDA with phone and data transmission features. None sold more than a few tens of thousands of units. Obviously, only the technology addicts were interested in using them.

3.1.2.7 It's a Mobile Phone, Stupid

Our development concept was, then, 'it's a mobile phone, stupid'. In external appearance, size, weight, price, talk time, and standby time, our chief objective was for the i-mode phone to be indistinguishable from existing mobile phones.

Of course, we wanted the display improved to make it easier to read all sorts of information. But we wanted the display enlarged only somewhat, because if most customers were not receptive to the new version, we would not get the numbers necessary to touch off a positive feedback process. Models with obviously different designs and ways of using them – a large liquid crystal display (LCD) with touch-screen features, for example – were ruled out from consideration.

Our 'it's a mobile phone, stupid' concept imposed a major restriction. By the time i-mode was in development, the shrinking of the Japanese mobile phone had gone quite far. Almost all manufacturers were offering phones that came in under 80 g and 80 cc in volume. Bigger phones would be a hard sell, no matter how much we claimed that the increase in size reflected heftier features. In our view, 100 g and 100 cc was the boundary point: if a phone was not smaller than that, consumers would reject it.

That meant that we had to squeeze the new features we wanted into a not-quite 100-g, 100-cc package, and we had to do that at an acceptable price point. That was no joke. At the initial stage, the problem was having enough memory to support the larger LCD and the wallpaper screen. Thanks to heroic efforts on the part of the manufacturers, we solved that one. But we could not upgrade the mobile phone's features all in one go. The reality was that we had to tailor the services offered to the evolutionary stage of the mobile phone.

3.1.2.8 Using Existing Web Servers

The other technical issue concerned the network system. That consists of the wireless network connecting subscribers and base stations, the gateway servers connecting to the Internet, and the applications servers operated

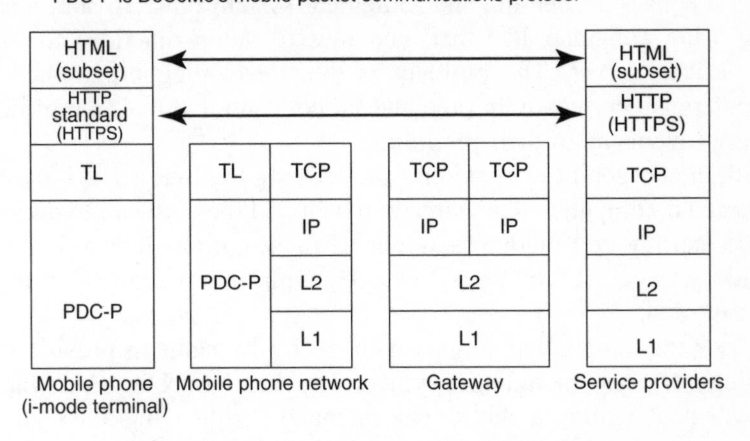

Figure 3.4 The protocol stack in the i-mode network.

by our service providers. The issue in building the network system was that it needed to take into account the characteristics of wireless transmission – slow speeds and high rates of transmission errors – while building a system that content providers would be able to work with easily.

Figure 3.4 shows the protocol stack for the i-mode network. It permits using standard Internet protocols for the application layer protocol handling interaction between the mobile phone and the server and a newly developed protocol tailored to the characteristics of wireless transmission for the transport layer. The aim was to keep adjustments made for the specific qualities of wireless transmission confined to only low layer protocols between the gateway service and the mobile phone.

The point of that decision was that content providers would have to make no changes in their servers whatsoever. That is, they could use the Web servers they already had for the Internet, unchanged, to provide content via i-mode.

3.1.2.9 Minimizing the Need to Change Systems

Our choice of Web servers was even more significant. In the conventional Internet, the role of most Web servers is that of a simple interface server. That is, all they do is create the graphical user interface for Web users. Behind them stand databases containing the information that people are seeking and applications servers handling the computation on the basis of that information.

An application server is something like the mainframe computer on which a bank's accounting and settlement system runs. To start a service using a big computer like that, you must develop functions for linking it to a Web server. The problem is that developing programs for big computers is very costly in time and money, and that makes it difficult to add improvements to their programs.

Still, the corporations providing services via the Internet to be accessed by personal computers have already developed the software to do so. Our goal in starting up i-mode was to enable those corporations to join us and provide service via i-mode at low cost, with only a minor development effort needed.

At present, i-mode has tie-ups with over 320 banks to provide mobile banking services, including payments by bank transfer – the Japanese equivalent of writing a check. The financial sector is the most conservative industry and the one that gives stability of service the highest priority. But i-mode was able to win the banks' support – precisely because we minimized the burden that making system changes to develop a service for i-mode would impose on them. And that was possible because we adopted the standard technologies of the Internet.

3.1.3 A Business Concept That Attracts Partners in Droves

We had settled our platform choice question on the basis of Internet thinking. But having a platform is not enough to get that positive feedback rolling. We needed to build an appropriate service design and business model.

The primary point we kept in mind in starting to design services was whether we could motivate potential content providers to leap at the opportunity to join us and deliver excellent services. In complex systems theory parlance, we were building the environment for emergence and self-organization. To that end, the most important task was to decide on the division of labor between DoCoMo, which provides the platform, and the content providers, which develop the services to run on it.

One possible choice in considering what DoCoMo's role would be was for DoCoMo to exercise its influence to the maximum degree possible. That is, DoCoMo might develop content itself, purchase content from others, or pay content providers a subsidy. Those are approaches taken by the mobile communications providers in Europe and the United States.

For the platform provider itself to start some sort of content business or to play a major role on content selection is, however, utterly contrary to the complex systems approach, in which we look for self-organization to start working.

If the platform provider starts its own content business, it is in competition with other content providers, and that will inevitably dampen its enthusiasm for developing its own content. That is similar to a situation in the software world: when Microsoft, which started out as an operating system developer, began developing the applications to run under its operating system as well, other applications software companies lost their enthusiasm.

If, however, the platform provider purchases content, then content providers would develop services that meet the platform provider's preferences. They would lose the will to develop content to meet customers' needs. It also is hard to imagine them staying engaged in the continuous development of better content. The results would not be very promising.

What if DoCoMo exercised no influence over i-mode content? The result would be like the situation on the Internet: too many Web sites, with too huge variations in quality. It would be hard for people to find the sort of Web sites they want, and that would be very inefficient – not an attractive situation for our subscribers.

3.2 DoCoMo's Role: Two Points Only

Having thought through those issues, how we defined DoCoMo's role as the platform developer boiled down to two points. First, DoCoMo will put its heart and soul into operating its portal site. Second, it will fulfill the role of guiding subscribers, who are attracted to it in the expectation of receiving some sort of service, to the appropriate Web site.

What is important about our role as guide is that we make it easy for subscribers to understand how the services are arranged. That is why we created our own i-mode 'content portfolio' and strategically allocated the Web sites to which DoCoMo has links according to the rubric of that portfolio (Figure 3.5).

In the i-mode content portfolio, all content is divided into four categories: information, e-commerce, databases, and entertainment. The 'information' category is for time-sensitive information – news and weather. The e-commerce category includes banking, securities transactions, and

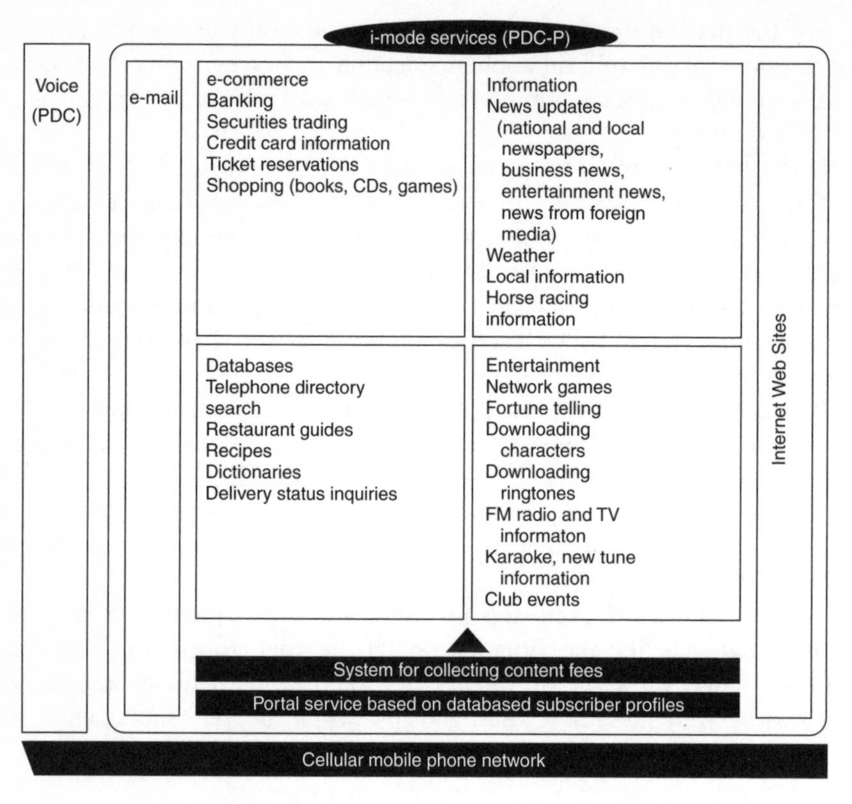

Figure 3.5 The i-mode content portfolio.

ticket reservations. The database category includes restaurant guides, dictionaries, and other information it is handy to have available in database form. The entertainment category offers services that are just for fun, including games and music for downloading.

By balancing content offerings in these four categories, we have made it possible for all subscribers to find the content they want and need. The point – and it is very important – is that subscribers know 'There's definitely content that matters to me.'

When I talk about i-mode overseas, I am often asked, 'What's the most popular content on i-mode?' or 'Which service is used most heavily?' It may be of interest to companies considering starting a service on i-mode to know which services are popular in general, but it is meaningless for individual subscribers. Regardless of what the big draws are, if i-mode does not have a service that the individual subscriber needs, there is no value in using it.

To DoCoMo, having chosen to function as a portal site operator, which content is most popular is also not very meaningful. Some content has few subscribers, yet attracts strong support, with subscribers constantly using it. Tiny publications that could not begin to turn a profit in print form can thrive on i-mode.

The idea of buying up content, which we rejected, would probably have entailed cutting off highly specialized content that attracts small but strong followings. Those asking 'What's the most popular content?' may be thinking in terms of a model of buying up content, of collecting popular content efficiently. That is thinking in terms of trying to keep profit-making businesses to oneself – a telecom way of thinking. So after explaining, as I do, why the question is not very meaningful, I add that it is not meaningful in an Internet way of thinking.

3.3 Sharing the Revenues Matters

Our commitment to performing our role as guide to the content available on i-mode is our stance with respect to our subscribers. Another key point is our stance with respect to content providers.

Our question was how to motivate content providers to join i-mode – and continue to develop better content. The answer we came up with was to collect their fees for them. DoCoMo bills our subscribers for their packet communications usage and for the content services they have signed up for. We retain 9% of the content services fees we collect as our commission for handling the collection and turn the remainder over to the content providers (Figure 3.6).

That is revenue sharing, an approach based on our concept of setting up a win–win relationship with content providers. We leave the content creation to the service providers who excel at that; DoCoMo concentrates on our system for collecting fees, our platform, and designing our data warehouse. That is what we might call our platform concept. DoCoMo's role, on which we focus, is to provide a win–win platform.

America Online (AOL), which grew so rapidly as an Internet service provider, is also based on a similar platform concept. AOL does not create content. It concentrates on designing the platform so that service providers are drawn in droves into the AOL community.

Because i-mode is an Internet-style business, we followed the same business model as AOL. Rather than buying content, we wanted to create a situation in which service providers can make money through DoCoMo's

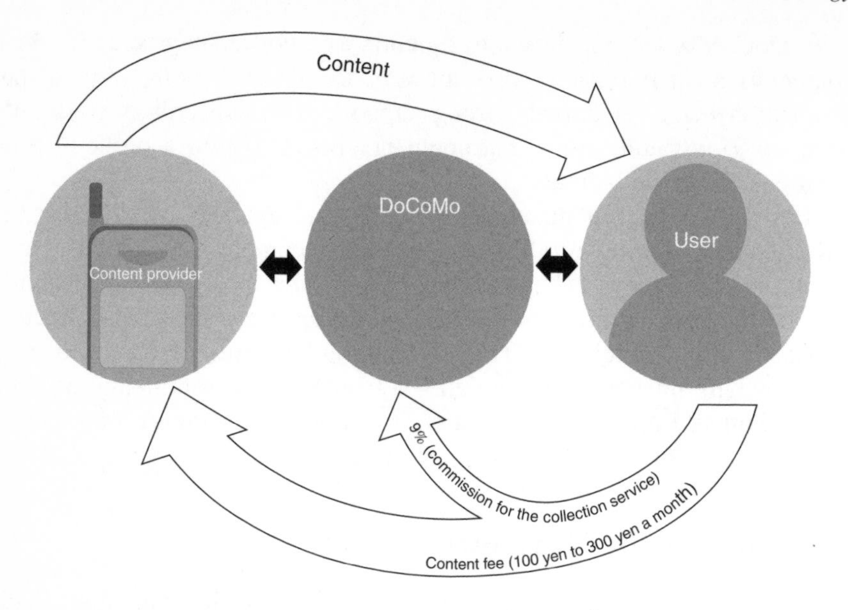

Figure 3.6 DoCoMo's fee collection service.

services. That is, we could work together and both would benefit. The rule we established was: 'DoCoMo makes money from our packet communications charges, while you make money by charging for your services.'

In this model, the more subscribers a content provider attracts, the more its revenues rise. But unless a provider continually renews its content, subscribers will not stay with them in the long run. That would motivate savvy service providers to develop attractive services and to keep refining those services after they launch them – or so we hoped.

3.4 Keep Service Providers Motivated

The telecom-thinking approach would be to provide not only the platform but everything else, including the content. Buying up content is another, closely related telecom-style notion.

DoCoMo actually is a company in which telecom habits of mind remain deeply rooted. We have our info-dial service: telephone numbers to dial for weather forecasts or horoscopes. We also have a mobile Internet service called *opera*. Both operate on the model of buying up content. That is a way of doing business that is probably shared by mobile phone companies anywhere.

In the process of working out what i-mode would be, a consulting company proposed the buying content model to us, at least for part of the i-mode content. Their argument was that Internet businesses are high-risk, high-return propositions; to minimize risk, we needed to launch with as much content on tap as we could. And if we bought up content and one of our content services proved to be a big hit, we would profit from it.

That model, however, would not help sustain service providers' motivation to keep improving their own content. At the point that their content was bought by a telecommunications provider, their content development would come to a halt.

3.4.1 Four Conditions for Attractive Content

What then should we do to keep subscribers using an i-mode service? From the subscriber's perspective, the ¥300 maximum monthly charge per content service adds up to ¥3600 a year – enough to buy a computer game or a video. And using content services on i-mode entails paying packet communications charges as well. We could not expect subscribers to pay good money – and keep on paying it – to access content on i-mode that they could find on the wired Internet for free.

What sort of content would subscribers happily pay service fees for? I think good content must satisfy four conditions.

One is freshness. That applies to all sites, not just to news sites, where the first with the story wins. You should be able to discover something new each time you access any site. If not, why would you keep visiting it? It is essential to build in a sense of the seasons, of time passing, and – depending on the nature of the site – perhaps to address current issues.

The second requirement is depth. The information must be detailed and in depth. It is not unnatural to think that because the i-mode screen is small, you can get away with providing shallow information. You cannot. Superficial information, information that pretends to be more than it is, turns subscribers off. Being a fiend for details is the way to go.

The third requirement is continuity. With an online service like i-mode, what is ideal is to have people using it every day, even briefly, and keep on doing so for an extended period of time. But no matter how fascinating the site, using it without limit is going to generate high packet communications charges in a short time. It is necessary to design the site so that subscribers have fun exploring its intriguing aspects and its depths little by little, in a long-lasting relationship.

The fourth requirement is a clear benefit: enhancing subscriber satisfaction. The subscriber should think that the site is fun, enjoyable, useful, and cool. The service must be designed so that subscribers always get some sort of satisfaction from it.

3.4.2 Nationwide Meetings Help Develop an Eye for Content

Keeping the quality of the content on the iMenu of the DoCoMo portal at the desired level is a nontrivial task. We realized that it would not do if the various people responsible for evaluating content were using different yardsticks – but we also realized that having a single person responsible for vetting all content was out of the question.

What is the solution? At DoCoMo, we pull together the people responsible for evaluating content at our regional companies nationwide for regularly scheduled editorial meetings. The purpose of these meetings is less to select content than to build shared editorial standards among the content managers.

At these meetings, content managers from each regional company explain content proposals, take questions from the group, and state their views. Because this process is used for all the proposals that each of the regional companies brings to the meeting, these editorial meetings are costly in time and money. They also generate huge amounts of reference material, since we make copies and hand out the hundred or so proposals. And, because nothing goes on the official menu without the unanimous consent of those attending the editorial meeting, considerable time is spent on each proposal. Lately, these nationwide meetings have been lasting two days.

At first glance, these meetings may look terribly wasteful. But as the participants go through the evaluation process again and again, what points to look for in checking content become clear, and the participants come to have the same standards for making judgments. This is a system that multiplies the number of content managers who can judge quality in the same way. Today, when a proposal is brought to one of our regional companies, the content manager there is able to judge it, using the same perspective and quality standards as at the nationwide editorial meetings. Some of these proposals may have been brought to the company's top management by someone with clout, but because all the content managers share the same thinking, that is not a worry. It does not matter whose proposal it is: the regional content manager can say, 'This is not good enough', and explain the reasons why.

The nationwide editorial meetings would be overwhelmed if they tried to process all the proposals that come in from around the country, but the system as it is working now is quite efficient, with the content managers at the regional companies doing the initial triage. Ultimately, I hope that we can develop our shared eye for content to the point that all the proposals the regional companies bring to the nationwide editorial meetings are approved.

3.5 What is Internet-Style Marketing?

In starting up i-mode, we selected the right technologies and the right business model. The remaining issue was advertising and marketing. What would marketing, Internet style, be? The answer is extremely simple: it is to push service, not technology. Our message is what you can do with i-mode, not why it is possible to do it. We have followed that approach consistently in advertising i-mode.

Today, when advertising for the wired Web, you will not see anyone announcing, 'Our service uses the Internet.' What do you see? 'Bank from home', or 'Make reservations from home with our service', with 'using the Internet' in small print. The message is structured from the perspective of the potential user of the service, not that of the service provider.

Why not, in Internet thinking, make the technology the main message? Because everyone is using it – it is the Internet. It is the *de facto* standard. And when you are using the *de facto* standard, you do not talk about technology. That is because you can talk all day about the technology and not tell anyone about the superiority of the service you are offering with it. And that is the real story.

There is, then, a good argument for saying that there is simply no need to make a big deal over the technology in Internet-style marketing. There is absolutely no value, at this point in time, in saying 'Our service is delivered using a browser' or 'We have introduced Internet access to our service.' Everyone is using the Internet. Everyone is using browsers. That is not news.

3.5.1 The Concept Behind the Commercial with Hirosue

We followed the above line of thought in developing the advertising strategy for i-mode. Not once did we say, 'Access the Internet from your mobile phone.' Nor did we use Internet terminology. Web, access, browser – our advertising is mum about them. For example, take a look at the first commercial we broadcast for i-mode (see color plates). The actress Ryoko Hirosue (who is known by her surname, Hirosue) is back stage, having her hair done, when she decides to make a bank transfer. She takes out her mobile phone and says to the hair stylist, 'Excuse me for just a moment.' He says, 'What, a phone call?' 'No,' Hirosue replies, 'a bank transfer.' That was the first advertisement for i-mode.

For the majority of people using their mobile phones to make a bank transfer – as commonplace a transaction in Japan for paying bills as is writing a check in the United States – it does not matter whether a banking service is based on the Internet. If we get across the message that it is possible to use banking services from your mobile phone, we have done our job.

Suppose we had decided to focus on the technology. 'We deliver banking from mobile phones, via the Internet; our highly secure service, using Secure Sockets Layer (SSL) technology.... ' Most people hearing that would think, 'Internet? SSL? What is all that? It's obviously too high tech for me to use', and flee. There was a good reason for our choice of nontechnical language and concepts in our marketing.

3.5.2 Who is the Lead Goose?

In marketing, at the stage of creating stimuli to touch off positive feedback loops, the vital question is: Who will be the lead goose in that self-organizing flock described in Chapter 2, in the discussion of complex systems theory? That is, we wanted to find the company or individual who, once in flight on i-mode, would inspire everyone else to follow. Identifying the potential lead goose, and getting it into the air, were our top priorities.

In the content domain, we saw our likely lead goose as one of the big nationwide banks. My tactic was to approach that bank first.

Making a bank our primary target was not an idea universally accepted at DoCoMo. 'Almost no one makes bank transfers or checks their balance daily. Surely there are other kinds of content – news or weather forecasts, say – that people would use more frequently?'

But I still decided that a bank would be first. Why? Quite simply, nothing is more conservative, more unwilling to break out of the pack, than the banking industry. And the nationwide banks are enormously influential. If one said yes, then the other nationwide banks would get on board too. And if they did, the second tier banks and regional banks would be sure to follow. And if the banks were on board, other industries would join them. 'If the ultraconservative banking sector is doing it', I could imagine people thinking, 'It might be worth thinking about whether there is anything in it for our industry too.' That was my plan.

My first step was to head for Sumitomo Bank, where I had an acquaintance from my previous job. While those outside the banking industry would not easily perceive Sumitomo's influence, it is the bank the others look to for leadership. My experience had taught me that if Sumitomo were to do something, the others would fall into step. I had another reason for picking Sumitomo as well: it had a more advanced computer system. A dominant bank that installs new systems: Sumitomo Bank was the ideal candidate to lead my flight of geese.

The Sumitomo, Sakura, and Sanwa banks were the first three corporations to agree to participate in i-mode. My plan had worked: with three big banks on board, other banks lined up to follow them. At the start of i-mode service, of the 67 corporations we had as partners in providing content, 21 were, in fact, banks – positive proof that my tactics were sound. It was a classic example of self-organization in a complex system.

A similar pattern applied to the media that cranked up the excitement about i-mode. As of the end of October 2000, 102 books have been written on how to use and enjoy i-mode. DoCoMo cooperates with their authors, but does not actively try to wangle that kind of coverage. The publishing companies planned those books on their own – with one exception. That was the very first of the i-mode books, for which DoCoMo dropped a strong hint to a publishing company that it might do a book. When that book sold well, other publishers hastened to follow. And not just to imitate the first book: competition naturally meant that each publisher tried to produce a better book than the previous one, target a different market segment, or otherwise add its own special touches. That is, self-organization occurred in the publishing of these 'how-to' books – since, fortunately for us, we had found our lead goose.

3.6 Finding i-mode Champions

As was the case for banks and in publishing, every industry has its leaders that can get the whole flight of geese to take off. Similarly, every organization has a key person who leads the discussion, and what that person says tends to go. Business schools call such individuals 'champions' and teach that, in the IT business, being able to find a champion is the key to success. If you can find one, then when you want to try something new, when you have hit a problem that must be solved immediately, you can go talk it over with your champion, and he will do something about it. When you are stuck, he is the fast source of help.

Companies that are followers in the IT world do not have champions. Or, rather, they have people who might become champions, but they do not give them appropriate positions. They do not make effective use of the resources they have on hand.

To put it another way, companies that do not succeed in IT businesses tend to be rigid organizations in which everything is done according to the book. They have got a manual to follow and, as long as they follow it, their employees are fungible. That is not a mode of thought that bodes well for success in the IT field. The manual cannot keep up with changing markets.

The passage of time in the Internet world has been likened to dog years: just as one human year is said to be equivalent to seven years of a dog's life, one year in the Internet world is equivalent to seven years in any other world. The pace of change is so fast that a decision made today, and three months from now, will have different meanings, even if the same decision is made. A champion must be someone who has the capacity to make decisions flexibly in response to changing circumstances.

The key to success in an IT business is to find champions outside your organization and to create champions within it who have the judgment to act in response to changing circumstances.

Chapter 4
Alliances

4.1 Win–Win Relationships

4.1.1 Sharing the Profits with Partners Who Share the Risk

Since the start of i-mode services was announced in November 1998, DoCoMo has announced a continuous stream of i-mode-related alliances.

It was only a few months later, in March 1999, that we formed an alliance with Sun Microsystems to cooperate in the use of Java technology. More recent alliances include those with Sony Computer Entertainment and Lawson Inc., a major convenience store chain, companies that at first glance might seem to have little connection to mobile phones (Figure 4.1).

In addition, while those relationships have not been formally announced as alliances, we continue to work closely with mobile phone manufacturers and service providers to promote i-mode services.

All these alliances are in line with the basic i-mode strategy. That strategy calls for making i-mode a core element in everyday life, used at all times, places, and occasions that people encounter in their everyday lives.

DoCoMo cannot, however, realize those goals on its own. We need not only the cooperation of manufacturers in improving mobile phone features but also the cooperation of firms involved in providing services that will be tied to people's daily lives. Each alliance with another company is another step toward our goal of making i-mode the core of everyday life.

Whatever the alliance, we strongly believe that it should be a winning proposition for our business partners. A business model in which a new business benefited only DoCoMo is one that no one would agree to

Year	Date	Alliance
1999	March 16	Memorandum of intent with Sun Microsystems concerning cooperation in use of Java technology
2000	March 29	Investment in Playstation.co.jp Investment in Web-based banker Japan Net Bank, Limited
	April 19	Investment in online payment service Payment First
	June 1	Establishment of D2 Communications joint venture with Dentsu Inc.
	August 1	Memorandum of intent to cooperate with Sony Computer Entertainment in technology development
	September 27	Contract with America Online (AOL) to develop and provide new Internet services
	October 5	Establishment of i-Convenience, Inc., a joint venture with Lawson, Matsushita, and Mitsubishi Corporation

Figure 4.1 i-mode alliances.

participate in, no matter how hard we tried to sell it. We must, we believe, offer a clear and demonstrable advantage to our partners.

That principle works both ways. While we receive many proposals for alliances, some appear to benefit only the would-be partner. We can see no advantage for DoCoMo and cannot imagine why we would want to accept these ideas.

We prefer to think up new businesses in cooperation with our partners and to share the profits with them. That is what we mean when we talk about win–win relationships. We make forming such relationships a basic rule in our strategy.

What kinds of benefits, then, do NTT and our partners derive from our relationships? Let us take a closer look.

4.2 Different Industries, Different Types of Alliances

Sorting out the various alliances that DoCoMo has formed so far, we can divide them into three types: *technology alliances* that promote the development of mobile phones and mobile phone services; *portal alliances* that promote the development of portals operated by DoCoMo; and *platform alliances* that expand the range of times, places and occasions in which mobile phones are used. Over time, technology and portal alliances evolve vertically; platform alliances increase the convenience and utility of i-mode horizontally by providing links with other platforms (Figure 4.2).

Our first and primary technology alliances were with phone manufacturers. Stimulating new and replacement demand is the benefit that manufacturers derive from adding new features to phones. When a telecommunication provider's introduction of new services is timed to coincide with the manufacturer's launch of the new types of phones required to use them, it stimulates demand from subscribers who now want these new phones. The provider's revenues also increase, since adding new applications increases traffic.

Mobile phone development has been a process of adding new features little by little. That is partly because the manufacturers were unable to make large additions to features all at once, while keeping the phones at the same price and size. Subscriber reluctance to accept radical change has been another major factor. New functionality in small increments is easier for users to swallow.

4.2.1 Technology Alliances for the Promotion of Mobile Phones

4.2.1.1 Music and Color Displays

Looking back over advances in i-mode phones, we find that the 501i series introduced at the start of service in February of 1999 added packet-data transmission and a HyperText Markup Language (HTML) browser to the conventional voice features of mobile phones, making it possible to browse the Web and to send and receive e-mail.

The 502i series introduced in December 1999 added support for downloading ringtone melodies. Technologically speaking, this was made possible by adding the same Music Instruments Digital Interface (MIDI) sound processing features used in karaoke-on-demand services. This innovation not only created a huge market for downloadable ringtones but also drove a sharp increase in sales of i-mode phones.

In addition to MIDI playback capabilities, some of the 502i series models were equipped with color displays. We had left the decision to add color displays up to the manufacturers. Color content was rare at that stage, and we were not at all confident of how such content would be accepted by subscribers. Thus, we decided to hold off from making color a standard feature.

By the second half of the year 2000, however, a wealth of color content had appeared. Even manufacturers who had been hesitant about color

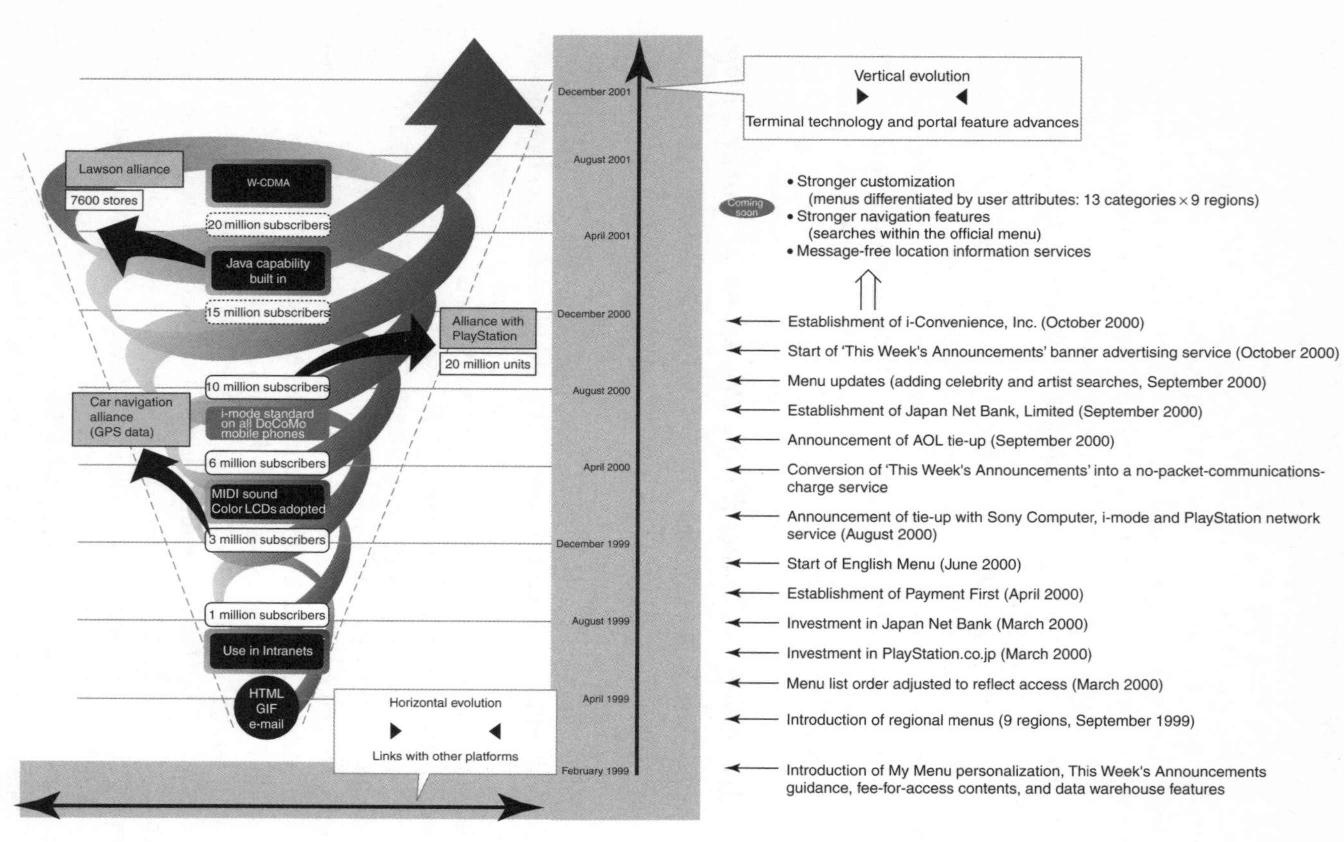

Figure 4.2 Two strategic axes, vertical and horizontal evolution.

began producing color-display phones. From this point on, virtually all new models were equipped with color displays.

The 503i series launched in December 2000 was equipped to handle the Java programming language developed by Sun Microsystems. Enabling phones to download programs written in Java made it possible to offer a wealth of new applications on them.

4.2.1.2 Openness Enables Personalized Phones

At the start of i-mode, I was determined to create an open platform for mobile telephones. From the user's point of view, that meant being able to customize a phone by selecting images and applications that fit personal interests. Previously, users who wished to personalize their mobile phones had been limited to choosing distinctive straps or stickers to attach to them. With i-mode, we made it possible to personalize the phone's insides – its software.

In conventional mobile phones, the manufacturers made the decisions about displays and applications to be included; everything was locked in when the phone was shipped. Users were left no room for choice. i-mode's openness made new choices available to them. As we moved from the 501i to the 502i and then to the 503i series, the course of i-mode phone development followed the path we laid out toward increasing programming openness supporting increasing personalization.

The first step, in the first-generation 501i series phones, was to open the display to the subscriber's control (Figure 4.3). The display could be adjusted to show the information best suited to the user's needs, and the subscriber could choose what image to display as wallpaper while in call-waiting mode. In the 502i series, we gave subscribers a choice of ringtone melody, further expanding their freedom. Now a subscriber could download the melody of his or her choice and give the phone a personalized ring.

The display with the 501i series, the ringtone with the 502i series, and then the 503i series opened up space in the phone's memory for downloadable applications. That is, space was set aside to store and run programs the subscriber had chosen to download. With that development, the mobile phone had become a computer, allowing programs to be added or changed. While previous i-mode phones kept personalization on the surface, that is, display and sound, this new step marked a radical breakthrough.

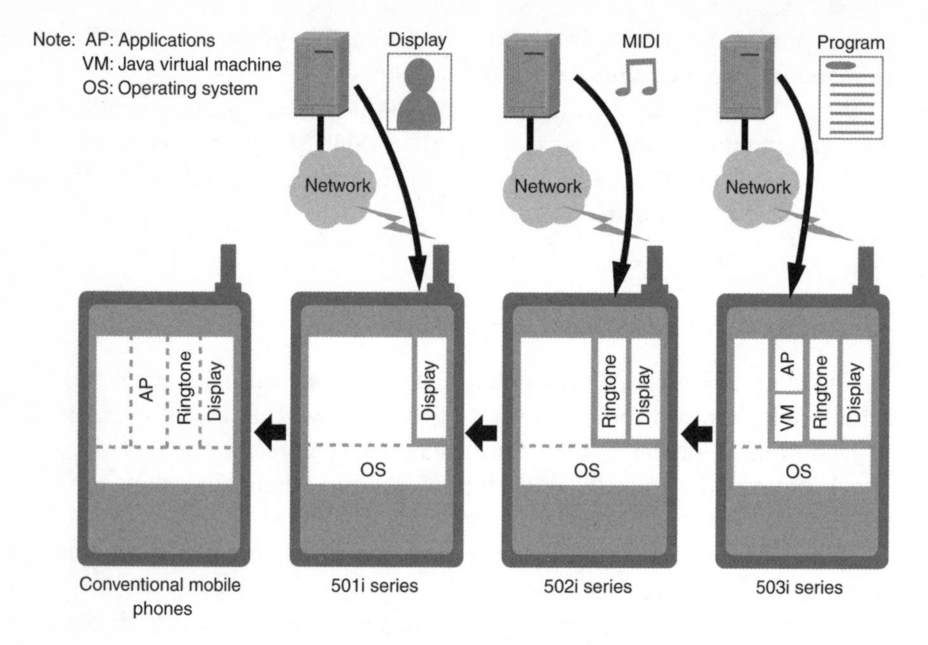

Figure 4.3 Opening up the mobile telephone.

4.2.1.3 Java: Continuing Use of De Facto *Standards*

The choice of the Java programming language for downloadable programs was consistent with our basic strategy of building i-mode on existing *de facto* standards.

When we considered the use of Java in 1998, several alternative languages for downloading programs were available. Companies such as Microsoft, which felt threatened by the rise of Java, had begun to propose their own technologies that were based on similar concepts.

At the end of the day, however, Java was the best choice, for there were already several million Java developers. We knew that if we did not choose the language that minimized barriers to entry for those who develop applications, we would be digging our own grave: when telecommunications providers began to offer this new service, there would be almost no applications ready to download. Since it was essential that our subscribers felt that being able to download programs benefited them at the start of service, not sometime in the future, we decided to go with Java, the language with the greatest number of developers.

Sun Microsystems also benefited by expanding the market for Java. Java had attracted considerable attention as a language for writing applications

"I never dreamed we'd be on the cover of *Business Week* this soon."

The Content Portfolio

E-commerce

Nomura Securities

Kinokuniya Bookstore

Today "mobile banking" has entered the general vocabulary. When i-mode was launched, 21 banks supported transactions by i-mode; as of October, 2000, over 320 do. The many other e-commerce sites offer airline and concert tickets, books, CDs, game software, and DVDs.

Information

WNI's Weather Information

Cybird's Mobile Map Service

Subscribers can access a host of information sources, including domestic and international news, weather, local information, and government information, with a few clicks of the button. Push-type information services also transmit information, handy when the subscriber is going out—news and sports updates, likelihood of rain, road conditions.

Databases

Sanseido's Sanseido Dictionary

Pia's i-mode Gourmet Pia

Search dictionaries, the yellow pages, restaurants, train routes—these tools solve many small problems. The restaurant search function is combined with map displays for even greater convenience. Banner ads and discount coupon screens add direct marketing functions.

Entertainment

Enix's Dragon Quest

Xing's Pokemelo Joysound

Bandai's Forever Kyarappa

In this field, the content on offer has mushroomed as i-mode has attracted more subscribers. Services transmitting cartoons and other images, ringtone melodies, and online games dominate, with the content becoming even richer as mobile phones have acquired color displays and MIDI playback features.

Magazine advertisement

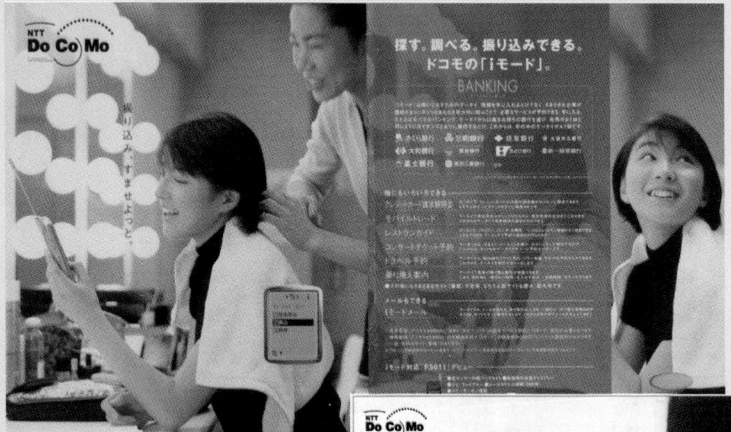

Newspaper advertisement

Marketing i-mode in a way consumers would understand. The advertisements used no IT terminology and stressed convenience for the user.

In-train hanging advertisement

Over 100 i-mode books.
(See Chapter 5, section 5.11 for details.)

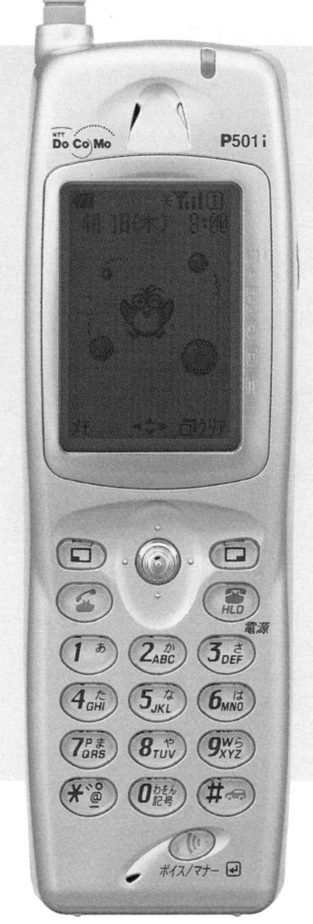

The 501i Series

The basic i-mode phone, designed towards future developments

The first i-mode phones looked like conventional mobile phones but had larger screens and memories and supported packet communications. The four-way navigation tool in the center and the "i" button with the logo were used on the first series and continued on the second series, the 502i.

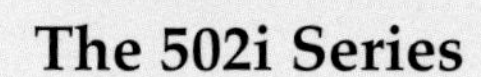

The 502i Series

Smaller, lighter, but with better visuals and sound

The basic functions were upgraded to store more e-mail messages, the phones are smaller and lighter, and they support i-melodies and i-animation. The handsets also come in more styles, some offering distinctive features such as infrared communication, connectivity to car navigation systems, and built-in games.

The 209i Series

By June 2000, standard models also were i-mode capable

Beginning with this series, launched in June 2000, i-mode capability was standard on the 20x series, our standard models. Their compact size and full features meant that replacement buyers would accept these phones.

The 821i Series

Super Doccimo combines PHS and i-mode to satisfy the demand for high-speed mobile communications.

These phones were designed so that subscribers could catch e-mail and news on the move with i-mode, then connect their handsets to a personal computer to access more detailed information at the 64 kbps the PHS standard supports. One device handles e-mail and telephone (mobile, PHS, extension phones).

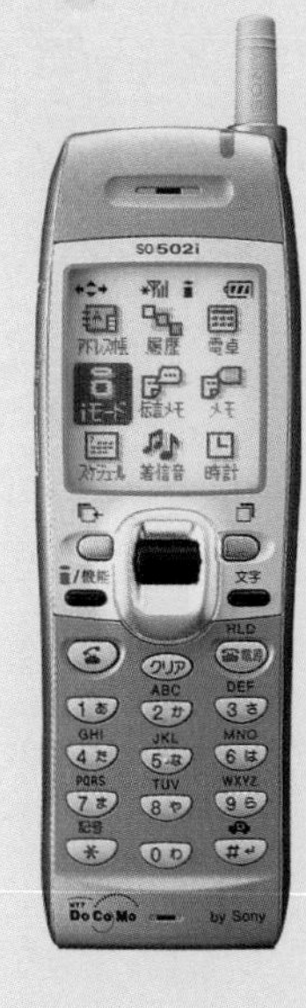

The 503i Series

The new series has Java capability built in—a huge surge in i-mode's capabilities. Downloading software gives access to more entertainment content, while beefed up security and user recognition functions give peace of mind for using i-mode for e-commerce.

And then, 503i, the next-generation model

W-CDMA Concept Models

DoCoMo launched the world's first W-CDMA service in May, 2001. High-speed packet communications (64 kbps uplink, 384 kbps downlink) make transmitting full motion video and large volumes of data go smoothly.

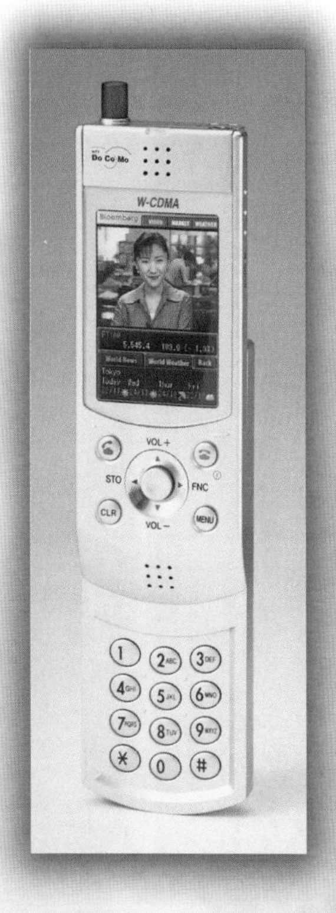

that could run on a wide variety of computers regardless of their operating systems. However, because of the overwhelming dominance of Windows in the personal computer market, there was a little felt need for a cross-platform language.

Thus, Sun's hopes had shifted to personal digital assistants (PDAs), set-top boxes for television sets, and other non-personal-computer applications. But while Sun understood the scale and potential of those new markets for Java, it did not know how to break into them. It was casting about for a way to establish a toehold. The timing of DoCoMo's proposal could not have been better for Sun: it carried Sun across that initial barrier into new markets. Sun had had no prior connection to the mobile phone field, but its relationship with DoCoMo provided an effective starting point for a drive to establish a powerful presence in that market.

Covering as it did a domain that neither company could master on its own, the tie-up between DoCoMo and Sun was truly a win–win situation. Sun CEO, Scott McNealy, was instantly taken with the i-mode phone, carried it with him everywhere, and introduced it in his speeches.

4.2.1.4 Calculating Backwards from Vision to Start of Negotiations

The start of negotiations with Sun was the critical event in forming the Java alliance. We first went to Sun Microsystems' headquarters in Silicon Valley to negotiate a Java license in December 1998. That, you will recall, was two months before the launch of the i-mode service (Figure 4.4).

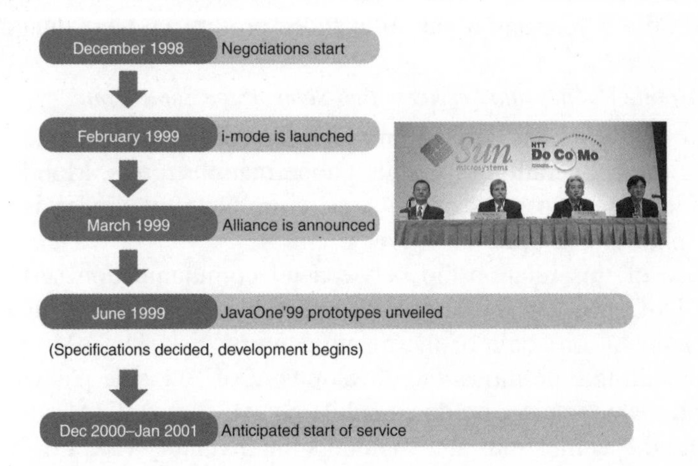

Figure 4.4 The alliance formation process with Sun Microsystems. Photograph courtesy of *Nikkei Microelectronics*.

We began negotiating a license for Java even before launching the i-mode service because we wanted to consider building Java capabilities into the third generation of i-mode phones. If that generation of phones were not Java-capable, delivering a whole host of additional services, including e-Money, would fall behind schedule.

A conventional approach to thinking about the timing of the next stages in a rollout would have meant waiting until we had accumulated a certain number of subscribers, or at least waiting until we had launched the service. It was clear to us, however, that that would be too late. Calculating backwards, we realized that if we did not allow ourselves at least two years, we would not be able to have Java-capable phones on sale in fiscal 2000.

Back then, Java had a track record in computers with powerful CPUs and lots of memory, but examples of its use in devices with very limited processing capacity and memory – such as mobile phones – were rare. Because it was necessary to define new specifications for use in mobile phones, we would, we realized, need development time.

DoCoMo and Sun Microsystems announced our alliance on March 14, 1999. That was barely three months after the start of our negotiations, an unusually brisk pace for setting up a business alliance, but the clear benefits to both parties had smoothed the way. Just three months later, in June, at the Java developers' conference, JavaOne '99, manufacturers displayed prototype phones that ran simple Java programs. We scheduled the beginning of Java-enabled commercial services for the end of 2000 or the start of 2001, only a year and a half after those prototypes were announced.

4.2.1.5 Mobile Phones and Services, Two Sides of the Same Coin

Before a mobile phone service provider can roll out new services, it has to secure the cooperation of mobile phone manufacturers. Mobile phones and services are two sides of the same coin. No new service is possible without phones equipped to deliver it.

Because of this relationship between telecommunication services and phones, DoCoMo and the phone manufacturers are mutually interdependent. By taking on some of the manufacturers' risk, DoCoMo preserves a relationship that promotes the development of 'saleable phones'.

Several manufacturers produce mobile phones, but DoCoMo determines the required specifications and shoulders the inventory risk. For DoCoMo mobile phones, we first decide on the specifications, then place orders for the initial lots with various manufacturers. Those lots are tiny, just enough for the manufacturer to break even. While a manufacturer will not suffer a loss by producing the initial first lot, it also will not make money doing it.

This approach frees manufacturers from inventory risk on the initial lot of the new phones. Even if the phones gather dust on store shelves, the manufacturer does not suffer a loss. The inventory risk for the initial lot is borne by DoCoMo. If, however, that initial lot does not sell out, that is it. We will not reorder those phones from that manufacturer. Manufacturers who hope to sell more than the first lot must produce good phones.

4.2.2 A Revenue Model that Encourages Service Providers

Technology tie-ups are one type of key alliance. Another equally vital alliance is the portal alliance. The content available on the portal site that DoCoMo operates, the iMenu, includes both fee-based services and free content. Service providers can benefit from providing either of them. The benefits to providers of paid-for content are easy to understand: the content itself is of value to our subscribers, and the service of providing that content is a business. Those fee-based services can be broadly divided into news services offered, for example, by newspaper publishers and entertainment sites offering downloadable ringtones and other enjoyable services (Figure 4.5).

Fee-based content businesses flourish on i-mode. While paid-for content is also available on the wired Internet, the content providers have been

Figure 4.5 The God of Love.

unable to attract a large numbers of users, so that these businesses are hard to grow.

In contrast, there are fee-based content businesses on i-mode with hundreds of thousands of subscribers and monthly revenues from several tens to several hundreds of million yen.

As described in Chapter 3, we made a critical decision that our business model would not include purchasing content. i-mode subscribers pay a monthly fee of between ¥100 and ¥300 for each of the fee-based services they choose to receive. After NTT's fee collection charge is deducted, more than 90% of that sum goes to the provider. That is, we have chosen to build a revenue-sharing relationship based on the assumption that the number of subscribers will grow.

Since, with this model, no subscribers means no income, a content provider cannot expect simply to start a service and watch the money roll in. To attract subscribers, a provider must, above all, create a distinctive service not available on the wired Internet or from other providers on i-mode. Having acquired a subscriber is, however, not enough. The provider must keep the subscriber interested; boredom with the site will prevent return visits and lose subscribers. At every stage, the burden on the provider is heavy.

When we were thinking through our content business model, some advocated buying content, arguing that otherwise the burden on content providers would be too great. What if their content did not attract enough subscribers? But that was the point: we were looking for providers who would be thinking ahead and coming up with all sorts of ideas to attract subscribers in the hope of future income.

Our choice of a revenue-sharing model turned out to be the correct one. All of our service providers redoubled their efforts to improve their content after they launched their services. They knew that more attractive services would attract greater numbers of subscribers and thus increase their income. And as their revenues grew, they could invest in the financial and human resources needed to add more value.

The *Nihon Keizai Shimbun* (the Nikkei, Japan's business daily) site is a good example. It started by offering quick access to breaking economic news, but it has built on that to add new services, such as breaking news on recent personnel movements. That is probably why this service has captured more than 90,000 paying subscribers. Since the site is priced at ¥300 per month, it generates an income stream of approximately ¥25 million each month for the Nikkei. With that much revenue, it is easy to think of adding another service on the site, especially given that adding a new service will further increase the number of subscribers.

Our structure thus provides positive feedback for individual providers as well as for i-mode as a whole. If i-mode were buying up content, we would not have seen services developing so vigorously.

4.2.2.1 Two Benefits from Free Sites

While the spotlight has been on the flourishing of fee-based content as a new type of business taking off on i-mode, there are, in fact, far more free sites on i-mode's official menu. As of the end of September 2000, of the 1160 items on the official i-mode menu, 800, or nearly two-thirds, were free.

While these sites do provide content free of charge, they still offer benefits to their providers. Two types of benefits quickly come to mind.

First, for any service business, mobile phones are a powerful means of keeping in touch with existing customers. Here we find banking, airplane reservations, and other transaction-based services (Figure 4.6).

The benefits to banks of using i-mode to provide such services as checking current balances and transferring funds lie in making it possible for customers to use banking services from their homes or other locations without making a trip to the bank. That increases customer satisfaction. Contact between banks and customers had previously been limited to the occasions when customers go to their banks. With the advent of access to banking services by mobile phone, banks now have many more points of contact with their customers.

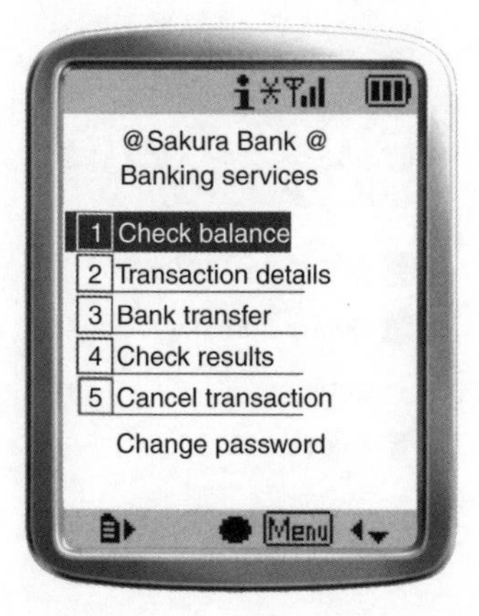

Figure 4.6 Sakura Bank.

The i-mode service also has potential for making existing bank operations more efficient. Since customers now are able to check their balances or transfer funds without having to travel to a bank, banks may be able to reduce the number of ATMs in their branches – or even reduce the number of branches.

Online banking via personal computers is also an opportunity to acquire new customers. However, given the limited spread of personal computers and Internet access in Japan, the potential for this form of contact is limited, and major increases in efficiency of operations have not been achieved. All the banks are now struggling with this problem.

The *sine qua non* for success is achieving a sufficiently large number of users. With the advent of a market with Internet access provided by mobile phones, online banking may finally attract enough users to achieve economies of scale. Some banks report that mobile banking customers now outnumber personal computer-based banking customers by more than two to one.

4.2.2.2 Service Providers Acquire Valuable Marketing Information

Another benefit of free sites is corporate publicity. That might be considered a form of advertising. Providing information via an Internet home page has many of the same objectives. Ajinomoto's *A-Dish* and Osaka Gas's *Bob and Angie* are recipe services that fall into this category (Figure 4.7).

Figure 4.7 Osaka Gas, *Bob and Angie*.

While neither of these companies adds revenue directly by using i-mode to provide recipes to their users, providing this kind of information adds to their images as firms involved in promoting dining culture. It does not matter whether the sites' users are Ajinomoto or Tokyo Gas customers. The important thing is the contact and the chance to firm up their brand images.

DoCoMo provides marketing data to content providers, regardless of whether their sites are fee-based or free. We provide data on such subscriber attributes as age and gender, together with usage patterns and the times of day they use i-mode.

Here again, the scale of our subscriber base is important, since achieving a certain sample size increases the value of statistical data. Content providers can use the data, not only for developing their sites but also in planning related to their bricks-and-mortar businesses' goods and services. That is another benefit we offer our providers.

The benefits are not, however, only for the service providers. Assembling a rich array of services and helping them grow has two types of benefits for DoCoMo. First, with useful and entertaining content, subscribers use i-mode more. Telecommunications traffic increases. And higher traffic volume means, of course, higher revenues.

Second, the more added value DoCoMo mobile phones offer, the more we can lock in subscribers. If we can add sufficient value that most subscribers feel, 'It would be so inconvenient not to have i-mode', or even, 'I can't imagine life without i-mode,' the less likely it is that they will switch to another carrier.

With penetration already in excess of 40% of Japan's total population, Japan's mobile telephone market is approaching saturation, making reducing churn and expanding market share an increasingly vital issue for mobile phone carriers. Without added value, the only mode of competition available would be a price war.

4.2.2.3 The Advertising Business, Data Customization Extension

Our advertising business is an extension of the portal alliances described above. To promote i-mode advertising, on June 1 2000, DoCoMo joined with Dentsu Inc. to found D2 Communications.

The key to success in the advertising business is having a critical mass of users, combined with a critical mass of advertisements: in other words, users who are targets for advertising combined with advertisers. Without both these elements, no advertising business can succeed.

D2 Communications was established in June 2000 because we anticipated that, by August, the number of i-mode subscribers would reach ten million. The opportunity to reach an audience of ten million was something that advertisers had long been waiting for. Since there were no precedents for mobile phone advertising, however, we needed a leading agency with the power to launch this new business. We decided on Dentsu, Japan's largest advertising agency. But while Dentsu is the prime mover controlling D2 Communications, D2 Communications itself functions as a media representative, a media-space broker. Thus, there is no barrier to its use by other agencies.

We might have set up a media representative by soliciting investments from several agencies. Since, however, mobile phone advertising was a totally unknown market, we needed people able to get it off the ground quickly. If several agencies had been involved, the staff seconded to join the new media representative would always be thinking of how its operations would affect the interests of the agencies from which they came. Making the necessary adjustments would take time. Luckily, Dentsu expressed a strong desire to participate. Besides 46% of the new company's capitalization, Dentsu also provided its president.

When trying to launch a new business, the shortest route to that goal is a business plan that includes a well-known business leader who will take the initiative in setting it up. If several companies had contributed only a few percent of the capital apiece, D2 Communications might have had sufficient funds, but it would have lacked that powerful leader.

On the Dentsu side, the benefit of this new business was not only a new source of income for the Dentsu Group but also an opportunity to launch a whole new advertising medium by itself. Dentsu had been a leader in TV, radio, newspaper, and magazine advertising. It would be an enormous benefit to the industry's largest agency to take the initiative in pioneering another new medium. Furthermore, to Dentsu, the world's largest agency, taking the lead in launching a new advertising market on mobile phones would be a highly significant achievement.

4.2.2.4 Portal Customization

One benefit of this new advertising business for DoCoMo was the strength it added to DoCoMo's i-mode portal service. As we explained while discussing the original directions for i-mode, at the stage when a certain volume of information has collected there, a portal service's role should be to guide users to the information they need. Some system is necessary to locate and bring to subscribers the information they really want from among the huge volume of content that i-mode makes available.

We are talking, in other words, about data customization. On i-mode, the first step in data customization was to customize the menus for each subscriber, starting with the 'This Week's Announcements' feature. We expect, little by little, to create menus that suit each subscriber's attributes, starting by changing to menus specific to the region of Japan where the customer lives.

Advertising is one of the ways in which people obtain the information they want. It is true that when we are looking for certain information, advertising may get in the way. Ideally, however, advertising should be a means for providing the information a user wants in the most appropriate way.

Our reason for not starting what would become a ten million subscriber advertising business earlier was our wanting to expand the medium to a large enough scale that both service providers and subscribers would spontaneously ask for advertising, and ask for it loudly.

The service providers were saying, 'We have a good service, but it's not reaching your subscribers.' With over 1000 sites on the official menu, the presence of each individual service is necessarily weaker.

All that DoCoMo can provide in the way of announcing new services is a mention in 'New Services This Week' when a new service was launched. We cannot keep reinforming subscribers about a particular service – or, indeed, about all the services.

We did expect subscribers to find the information overload frustrating: 'There is too much information, I can't find what I need.' Even with the official menu broken down into categories, when the total menu reached more than a thousand items, each category would still include several hundred items. The particular information a subscriber wanted could easily be lost in the crowd.

Advertising was one way to solve both sets of problems. Providers could pay for advertising scheduled to reach users when they had something to sell. Users would be able to click once to receive information they wanted, but would also be free not to click. So they would not feel under pressure.

If we could create a thriving advertising business, it would, we believed, strengthen our portal service. We waited, then, until the need for advertising was widely recognized before starting to offer i-mode advertising.

We have already started using the i-mode display for advertising and found that the maximum click rate for banner ads on i-mode was, on an average, 22%. That, as people who know the wired-Internet advertising business will recognize, is a very high click rate. This makes i-mode a highly efficient way for service providers to communicate with subscribers.

4.2.2.5 Advertising Tailored to Hobbies and Interests is Valuable Information

From the launch of i-mode, we already had in place an interesting scheme for portal development: it was what we called 'Message Free.' If a subscriber selects 'Options' from the i-mode top menu, she will see the message service selection menu. All it says about 'Message Free' is, 'This service, to begin in the near future, brings you information free of packet-communications charges.' We have made no special effort to advertise this service, but as of the end of September 2000, 1.2 million subscribers had turned it on. In other words, there were more than a million people waiting for information of some sort. As of the same period, more than 500 000 subscribers had used the options menu to register such personal attributes as birthdays and special interests (Figure 4.8).

Our intention is to use the registered data as the base for starting new information-distribution services. For example, we could satisfy a request from a service provider to transmit a certain message to 25-year-old women living in Shinagawa Ward in Tokyo.

Using this scheme, service providers can transmit messages efficiently to selected target audiences. And by registering their attributes, subscribers can receive information of particular benefit to themselves. Even though it is advertising, it fits the subscriber's interests, and thus to that subscriber it is valuable information.

While DoCoMo supplies the platform for matching advertiser needs and target attributes, D2 Communications operates as a media representative, securing advertising to use the platform. Naturally, because of privacy

Figure 4.8 My Data Registration on Message Free.

issues, subscriber information is never disclosed, either to advertisers or to D2 Communications.

4.2.2.6 Our Net-Only Bank is the Leader of the Pack

In addition to the examples described above, Japan Net Bank Limited (JNB), for which DoCoMo supplied 5% of the capital, can also be classified as a portal alliance.

JNB is Japan's first Internet-only bank. In forming the new bank, Sakura Bank and Sumitomo Bank took the lead. Besides DoCoMo, other investors included Fujitsu and Nippon Life Insurance. Service began in October 2000, and i-mode access to the bank is to begin in November.

In the future, the majority of financial services are likely to be provided via the Internet. JNB is the pioneer in providing financial services using this technology. From DoCoMo's perspective, JNB is likely to be the lead goose as in the complex systems theory model described in Chapter 2, and that is the reason for our investment.

JNB offers a higher level of service, including deposit notification by e-mail, than do the mobile banking services offered by other banks. It is able to do this because it built its backbone systems specifically to use the networks; it designed those systems with remarkable flexibility for adopting new services. We believe that with the advent of JNB's advanced services, ordinary banks now have an incentive to improve their own online services.

The benefit to JNB is that its clients can access its advanced financial services via mobile phones as well as computers. To DoCoMo, the emergence of financial services provided exclusively through the Internet is another opportunity to demonstrate the convenience of i-mode to its users.

4.2.3 i-mode Makes Daily Life More Convenient, More Fun, More Affluent

Besides technology and portal alliances, i-mode is also involved in a third type of alliance – platform alliances. Platform alliances are intended to expand the range of settings in which i-mode can be used (Figure 4.9).

Our long-term goal is, by making i-mode the core of everyday life, to make our subscribers' lives more convenient, more fun, and more affluent. The significance of expanding opportunities to use i-mode lies in bringing us closer to this objective.

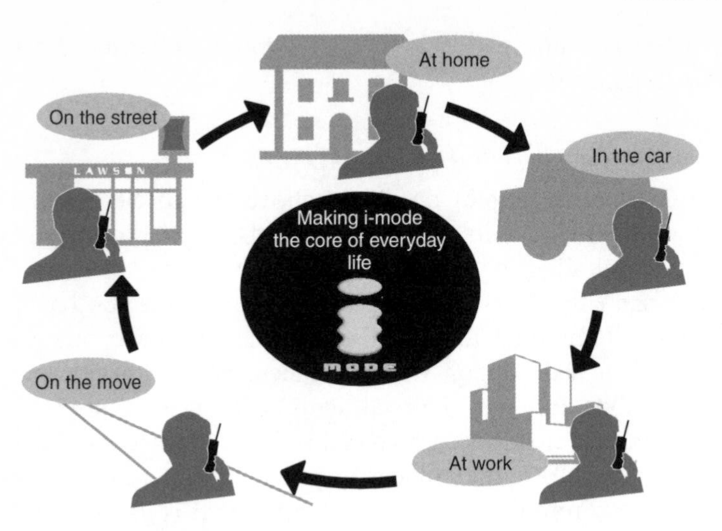

Figure 4.9 Expanding opportunities to use i-mode.

To make i-mode the core of everyday life, we want people to use it at all times, places, and occasions that make up everyday life. This is the basic concept that underlies the formation of platform alliances.

The first possibility that we looked into was using i-mode in cars. Since people in cars are away from home or work, the affinity between mobile telephones and cars is strong. Cars also offered a way to overcome i-mode's chief weakness – its small display. I refer to car navigation systems. From i-mode's perspective, what we wanted to do was to use the large car navigation display. Since people had already installed this nice, big display in their cars, we thought it would excellent if we could use it.

While car navigation systems were designed to display maps, there was no obstacle to their displaying i-mode data. With a mobile phone hooked up to the car navigation system, its screen could display i-mode menus, expanding the range of applications for which car navigation systems could be used.

Even more interesting, combining a car's navigation Global Positioning System (GPS) positioning data with i-mode information services could make possible services with higher added value. What kind of services might they be? We could, for example, use positioning data in retrieving other information.

With car navigation systems linked to i-mode, service providers can access GPS data on subscribers' cars. The provider's server requests the subscriber's current positioning data (latitude and longitude) from the navigation system. The car navigation browser returns the requested data. The server uses the positioning data and HTML's Common Gateway Interface (CGI) to search its database for information, which is then sent back to the subscriber.

Using this scheme you could, for example, display a map giving all the Japanese-style restaurants located within 300 meters of the subscriber's current position. We already support services using positioning data such as Townpage (business telephone numbers), iMapFan (maps), and Zagat Tokyo (a restaurant guide) (Figure 4.10).

We make every effort to minimize the burden on the server by using i-mode's own servers instead of Townpage or iMapFan's dedicated servers. When our server receives the http header, it recognizes that it is connected to a car navigation system and returns a dedicated car navigation display file.

We began forming car navigation alliances in March 2000. We were not selective in our choice of partner companies, but Matsushita Communication Industrial Co. Ltd., a firm in both the mobile phone and car navigation businesses, was the first to put an i-mode car navigation product on sale. Other car navigation system suppliers soon followed suit.

While our car navigation businesses have not yet grown very large, it remains a very big step toward expanding the settings in which i-mode is used.

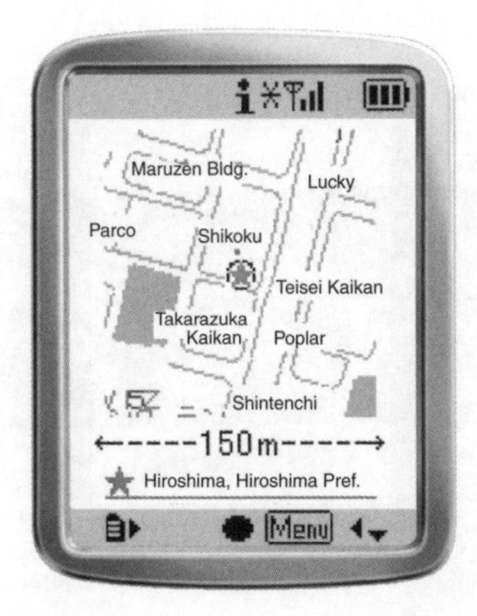

Figure 4.10 iMapFan map provided by Increment P/Mitsui & Co., Ltd.

4.2.3.1 PlayStation Alliance Promotes Home Use of i-mode

Our next step in forming platform alliances was a tie-up with Sony Computer Entertainment Inc. (SCEI). On August 1, 2000, DoCoMo and SCEI announced that they would develop new services linking i-mode and PlayStation, the Sony game machine. This tie-up became the core of our alliance strategy to make the home another setting for i-mode use (Figure 4.11).

As in the case of car navigation systems, the immensely larger TV screen provided another opportunity to compensate for the limitations of the i-mode display. Some other computers are able to control the TV screen, but none had achieved penetration to equal PlayStation's capability. Looking to the future, set-top boxes for digital broadcasts might be another way to reach the home market, but, at the time the decision was made, PlayStation was already there with the lion's share of the home video-game market.

Naturally, there were benefits for PlayStation. An alliance with i-mode would expand the time available to play PlayStation games beyond time spent at home. In addition to games provided on CD-ROM, new games to play on mobile phones could be provided by the PlayStation-Alpha service.

An example demonstrated when DoCoMo and SCEI announced our alliance was a Sumo game. While away from home, a player can use his mobile phone to train his wrestler and feed him the Sumo wrestler's favorite *chanko* stew. Returning home, the player can then have his wrestler fight, watching the match on the TV screen (Figure 4.12).

Figure 4.11 Tie-up with Sony Computer Entertainment. Photograph: Takanari Yagyu.

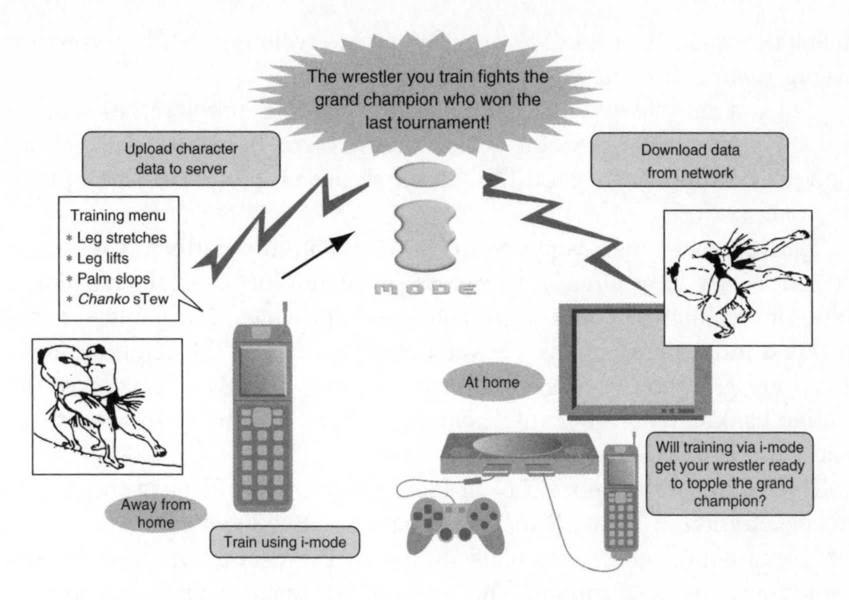

Figure 4.12 Content via the i-mode and PlayStation alliance.

It is now possible to experience this seamless enjoyment of a game that can be played both at home and away from home. With a Java-capable phone, it will be possible to download more complex games. Make no mistake about it – new, more enjoyable games will soon make their appearance.

We can imagine scenarios in which a command sent from a mobile phone to a server results in a response from the server that causes a hidden character to suddenly appear on the phone's display. It is not very practical to download the character data, but downloading the key that unlocks data stored on a CD-ROM would be a different story. In this system, the key makes it possible to see the hidden character.

4.2.3.2 Game Machines – The Most Familiar Form of Access

Online games of the type described above are not a new game genre. It is already possible to play online games over the wired Internet. However, in most homes the telephone jack is not located near the TV. In addition, now that Internet usage is becoming more common, the number of households needing two lines is likely to increase; it used to be that one telephone line per family was enough. However, not that many households have Internet connections, and, if you think of what is involved in hooking up a game machine attached to the TV to the Internet – running cables through the living room – it all seems a bit clumsy.

Using a mobile phone eliminates the cabling problem. In addition, packet communications, for which charges are based on data volume, and not on time, are a good fit with the desire of gamers to enjoy playing for long periods.

The reason i-mode was chosen as PlayStation's partner was because it had a user base already in excess of 14 million subscribers, plus the ease of the Internet connection that i-mode provides. For gamers, i-mode offers a more familiar link between PlayStation and the Internet than do fixed-line telephones. Since, moreover, i-mode already offers a wealth of online banking and other information services, the ability to access those services via PlayStation is also attractive.

To DoCoMo, the appeal lay in PlayStation's more than twenty million set penetration in Japan. If mobile phones are now the most common form of telecommunication terminals in Japan, PlayStation is now the most widespread home terminal. That is why we made a vigorous approach to SCEI.

4.2.3.3 A Packet and Storage Strategy That Makes i-mode a Home Entertainment Platform

Following the PlayStation tie-up, our next step was to expand nongame services. The first nongame service we thought of was catalog shopping. If catalogs that are now printed on paper could be distributed on CD-ROM or DVD and viewed using a PlayStation, there would be a huge benefit to catalog-shopping companies. It is extremely costly to print and distribute paper catalogs, but CD-ROMs and DVDs are inexpensive. The cost of burning and distributing a single CD-ROM has fallen to around ¥200.

Moreover, connecting a PlayStation to the Internet adds value to the CD-ROM or DVD catalog. By combining the catalog with an electronic key transmitted over the network, it is also possible to reduce the number of catalogs sent to the same address. First, a CD-ROM or DVD catalog with half-year's worth of data is sent to a user. Then, on a fixed date once each month, a key is sent via the network that opens that month's offers. Thus, for example, a key sent on December 1 reveals special offers for Christmas – offers that remained hidden until November had passed.

Use of this technique is not confined to catalog shopping. It can also be used by the travel industry, where the content and pricing of offers is highly seasonal. This combination of CD-ROM (the storage medium) with the key (transmitted by packet communications) is ideal for all sorts of catalog businesses. We call the creation of such services our Packet & Storage strategy.

However technologically advanced broadband transmission becomes, the cost of transmitting data will be high compared with other ways of getting information to people. Downloading the huge volume of data contained in a catalog will continue to be costly in terms of both time and money. It is simply not very feasible. That is what makes distributing the data in advance, whether in CD-ROM or in some other storage medium, a superior solution. It eliminates the need for multiple downloads and keeps data transmission costs miniscule. If alternatives to download are available, they will be preferred. This, in a nutshell, is our Packet & Storage solution.

4.2.3.4 PlayStation Users Continue to Increase

To PlayStation, the primary benefit of these developments is increasing the amount of time that users spend with PlayStation. While at present kids and young people are the bulk of PlayStation users, the addition of catalog shopping will expand PlayStation's potential user base to include other market segments, starting with older women.

At present, mothers are among PlayStation's opponents: 'Stop that and study' or 'Just one more hour', they say. However, if PlayStation becomes a medium for catalog shopping, they, too, will become PlayStation users. That means that contact with older women, a segment that has seemed to have no connection with PlayStation, will be established. Given the addition of shopping as a killer application, we anticipate that both the number of PlayStation users and the length of time PlayStations are used will grow.

This scheme also has benefits for catalog-shopping companies. These companies are already involved in producing electronic versions of their catalogs, adding product information displayed on Internet homepages to the printed catalog that they mail to their customers. However, their homepages receive very little usage.

It seems difficult to establish a comfortable link between personal computers and catalog shopping. Most people pick up a catalog when they feel like it and relax as they look through it. To use a computer to do that, they have to turn on the computer, connect with the Internet, and find the right Web server. This is a much more complicated and time-consuming process than reading a conventional catalog. And, in Japan, communication costs while perusing catalogs online make it impossible to relax.

In this respect, TV is the closest equivalent to a catalog. Since, moreover, each PlayStation is, in fact, a powerful computer, doing keyword searches for desired products will be simple. That is why we believe that i-mode combined with PlayStation is the ideal platform for e-commerce in the home.

4.3 Alliances with Broadcasters

As an extension of our Packet & Storage strategy, we are also considering alliances with broadcasters. Since full-fledged digital broadcasting may become a substitute for storage on CD-ROM or DVD, we see these alliances as part of this strategy.

Within a few years, broadcasters will switch from analog to digital broadcasting technology. Japan has two types of satellite broadcasting, communications satellites (CS) and broadcasting satellites (BS). CS digital broadcasting is already in operation. December 2000 saw the start of BS digital broadcasting as well. Next-generation (110° East Longitude) CS technology, which will allow the same receivers to receive BS as well as CS digital broadcasts, is scheduled to begin in the second half of 2001. Terrestrial digital broadcasting is expected to begin in 2003.

Our Packet & Storage strategy using CD-ROM or DVD uses our network to transmit a digital key that unlocks already distributed content. Given that digital broadcasting is a one-way 'downpour' medium in which high volumes of digital data are transmitted to the user, would not it be possible to include not only the key but also the stored component in the downpour? If next-generation CS broadcast receivers are equipped with hard disks capable of holding tens of gigabytes, could not they too be used as a storage medium?

The per-bit cost of digital broadcasting makes it by far the cheapest form of data transmission. It is possible that BS and terrestrial digital broadcasts will be distributed to households for free. To restrict the use of telecommunications circuits to placing orders and to distributing data by broadcast – that is the right combination.

For example, suppose that there is a video you want to see. We can imagine a service such that when you use telecommunications circuits to place your order, the video is automatically stored on your digital broadcast receiver's hard disk within the next twelve hours. The video would actually be transmitted when the orders for that video top 1000, in a way that allows only those who have ordered it to see it. This could be an extension of the Pay-Per-View system already used by CS broadcasters, allowing centralized control of who is allowed to receive the data. The combination of broadcast media that lack uplink signals with two-way i-mode packet communications will produce an extremely attractive medium. The asynchronous nature of broadcast media is not inherently a bad thing. The question is how best to combine it with other forms of data transmission.

4.4 Seamless Links between Mobile and Fixed-Line Networks

It probably will be, in a way, different from what we envision for our tie-up with PlayStation, but alliances involving the wired Internet are another way to expand home use of i-mode. For our wired-Internet alliance, our choice of a partner was America Online (AOL), the world's largest Internet service provider (ISP). On September 27, 2000, DoCoMo and AOL announced their new tie-up.

The alliance framework comprises three key points: (i) DoCoMo became the leading shareholder in AOL Japan and participates in its management; (ii) the companies work together to develop and promote Fixed Mobile Convergence (FMC), services combining fixed-line and mobile services; and (iii) they will invest in companies with the technology needed to realize FMC. Looking to the future, they aim to provide FMC services worldwide.

The alliance's first project was to provide free AOL e-mail accounts to i-mode subscribers. That is, all i-mode subscribers would have the right to sign up for a free AOL e-mail account, allowing them to receive e-mail via their computers' Web browsers. Both corporate account users accessing the Internet through company intranets and subscribers to other ISPs would be able to read their mail from an AOL Web page with no sign-up fee.

At this stage, we made full use of the AOL servers' filtering capabilities to forward e-mail to i-mode addresses. This allowed us, for example, to forward only company mail to an i-mode phone or restrict forwarding depending on mail content, through use of other fine-grained filters.

According to surveys of random samples of i-mode subscribers, many do not access the Internet via a personal computer or have an account with an ISP. By providing free e-mail addresses to such subscribers, we hope to be able to get them to sign up with AOL.

In other words, the benefit to AOL of this alliance is being able to use free e-mail addresses as an inducement and thus increase the number of AOL subscribers. The greatest single cost to ISPs is the cost of recruiting subscribers, an amount said to range between five and ten thousand yen per new subscriber. This alliance should enable AOL to sharply reduce these costs.

4.5 Future Plans Call for Adding PC-Internet Users

There are huge advantages for users in linking the wired Internet and i-mode. Having separate DoCoMo e-mail (@docomo.ne.jp) and AOL e-mail (@aol.com) accounts allows more intelligent use of e-mail. For example, mail to the DoCoMo account can be restricted to certain important people or emergency messages, with other mail sent to the AOL account.

The immediate objective of the alliance with AOL is to boost the number of subscribers to AOL Japan to more than a million. Once our one-million subscriber target has been achieved, we can think about full-fledged FMC services: for example, shared content or e-commerce tie-ups.

As I see it, we are going to experience major growth in personal computers and Internet access in Japan. Since i-mode already has more than fourteen million subscribers, a huge population already knows the benefits of network-based services. If we look closely at data concerning i-mode subscribers who have tried the Internet or network-based services, their top demands are for larger displays and greater convenience. Thus, while we often hear people talking about computers and mobile phones as if they were rivals in the networking market, in Japan the mobile phone may become an inducement for greater use of computers. While the spread of computers and the Internet remains lower in Japan than in the United States, the explosive growth of i-mode suggests high potential for similar growth in the wired-Internet segment as well.

4.6 Point & Mobile Strategy and Convenience Store Alliances

Considering new occasions for i-mode use, we thought first of cars (car navigation systems) and then of the home (PlayStation). Our next possibility was fixed street locations (service points).

Convenience stores are our largest single target for service-point applications. The first fruit of this project was the announcement, on October 5, 2000, of an alliance between DoCoMo, Lawson, Matsushita Electric Industrial, and Mitsubishi Corporation to create a company named i-Convenience, Inc.

Lawson, which has a nationwide chain of stores, and DoCoMo have joined hands because convenience stores and i-mode complement each other, that is, compensate for each other's weak points. The mobile phone is available wherever its user goes but provides no place to pick up products purchased online. Convenience stores have several thousand service points nationwide but have no way to track customer behavior once the customer steps outside the store.

Combining mobile phones and convenience stores makes it possible to offer seamless services (Figure 4.13). We could, we believed, create a world where the combination of the mobile phone and the convenience store would mean a trouble-free life. This is what we labeled our 'Point & Mobile' strategy.

To DoCoMo, the benefit of this alliance lay in increasing subscriber convenience by providing connections with physical distribution points. For example, it would be possible to pick up tickets ordered via i-mode at a Lawson convenience store. Even with an i-mode virtual shop as a place to make reservations, there still has to be a place, somewhere, to pick up the physical tickets. While DoCoMo does have mobile phone

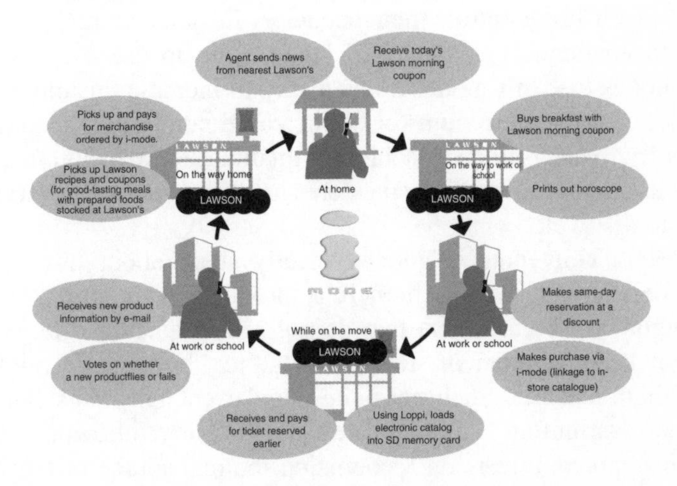

Figure 4.13 i-mode plus convenience store service concept. *Source*: i-Convenience.

retail outlets all over Japan, customer satisfaction with the service would be much higher if we used the convenience stores, points more closely linked to customers' everyday lives.

At present, convenience store staff must still issue tickets manually. In the future, however, this process will be automated. The service works like this: when a ticket reservation has been completed via i-mode, the subscriber receives a reservation number. When the subscriber enters the reservation number into a Loppi (a multimedia kiosk in a Lawson store), the ticket is issued. In the next stage, an IrDA standard infrared interface or another interface now under development, for example Bluetooth, may be used to communicate the number to the terminal. That will eliminate the time and trouble required to enter the number manually.

The next step after issuing tickets will be links with point of sales (POS) cash registers so that people can use their mobile phones to pay for convenience store purchases.

DoCoMo cannot, however, create these new businesses by itself, because it will take more than adding features to mobile phones. It also requires changes to in-store multimedia terminals and POS cash register systems.

4.6.1 Convenience Stores Benefit by Staying in Touch with Customers

To Lawson, the benefit is an increase in contact with customers. Lawson's 7600 stores located throughout Japan attract eight million customers per day. But as soon as these customers leave the stores, Lawson has no way to track, much less control, their behavior. In other words, Lawson has no way to guarantee that they will come again to use a Lawson store. Lawson hopes that in i-mode, it has a way to increase customer visits.

The factor that will account for an increased number of customer visits is the ability to use i-mode to achieve a qualitative expansion in the range of goods and services that Lawson stores handle – goods and services that are sure to sell well.

At present, convenience stores typically stock about three thousand items. There are, however, numerous products with strong appeal to limited numbers of customers that do not sell in quantities large enough to warrant keeping them in stock at all stores, but i-mode might make offering them feasible. Such not-quite-popular-enough items can be kept in regional distribution centers for next-day delivery following an i-mode order. This approach rests on recognizing the real source of convenience stores' strength. The sheer number of convenience stores has often attracted attention, but their real strength is their physical distribution networks and the outstanding information systems that provide support for their business.

New products can be held at distribution centers to limit inventory risk. Examples include cosmetics, nutritional supplements, vitamins, and similar products for which customers have strong brand preferences. While there may not be enough customers to warrant stocking these products at every store, they still can be stocked at distribution centers at a reasonable cost. The result will be to expand convenience stores' merchandising potential.

Using this scheme, Lawson's distribution network could support high-volume sales of products other than the three thousand products stocked at stores. Since delivery trucks haul products from distribution centers to every store twice a day as it is, the impact on delivery costs of such an increase in the range of products handled would be minimal.

Most customers stop by convenience stores either on the way home from work or during their noon breaks. In most cases the convenience stores they use are those close to home or those close to work. All they need to do is preregister those two stores with their i-mode phones, and they will be able to pick up the products they want within twelve hours of placing the order.

4.7 Printing: Our Next Project

As part of our alliance with Lawson, we are now thinking of a new project utilizing the high-performance printers installed at Lawson convenience stores.

There is a limit to how much information can be viewed comfortably on an i-mode phone display. By meeting the needs of i-mode subscribers who want to look more closely at the information they have discovered via i-mode searches or would like to have it printed out on paper, we need some other solution. There is clearly no guarantee that each household will have a computer and color printer, and it hardly makes sense to tell our subscribers that they have to buy computers and color printers to output information located via their mobile phones.

That is why we are thinking that Lawson, with numerous stores nationwide, offers the solution we need. Lawson already has high-performance color printers/copiers installed at its stores. Why not use them?

What would people want to print out? Travel-related pamphlets would be ideal. Yes, it is possible to find travel information on the Web, but reading it on the computer screen leaves something to be desired. When planning a trip with family, friends, or loved ones, you want a pamphlet with beautiful pictures to look at while you are talking over your plans.

'For our next vacation, how about a week in Amanpulo?'

'I don't know about the hotel rooms and restaurants.'

'OK, let's go to Lawson's and print out a pamphlet.'

That is the sort of use we have in mind. Since travel agencies already have data available in HTML on the Internet, all it takes to print out a pamphlet is using i-mode to specify the file location and the store at which you want to have it printed.

The benefits for travel agencies that put together tours are also enormous. Production costs for travel-related pamphlets are high, and the question of how to provide materials to those planning trips more efficiently is a constant headache. A printing business tie-up between DoCoMo and Lawson could help solve this problem.

Besides travel pamphlets, maps for trip destinations, stock market charts, and discount coupons for entertainment facilities are additional possibilities.

From Lawson's perspective, the printers in their stores would get more usage and become a new income stream. Who pays for the printing, the user or the information provider? That can depend on what is printed. For example, subscribers might pay for stock market information, while travel agencies pay for travel pamphlets.

As an extension of this business, Lawson might want to get into overseas mail order. Music downloads are another possibility. The possibilities are endless – but we cannot start them all at once. We want to proceed step by step.

While considering a convenience store alliance, we talked to several convenience store chains. Lawson was the quickest to understand our Point & Mobile strategy and display a real eagerness to help us implement it. DoCoMo does not intend to work exclusively with Lawson and would like to get all the convenience store chains involved in this business. But Lawson, by announcing its intention to set up a new company and move quickly as our partner to launch these new services, has become a leader in the market.

As in the case of the advertising business described previously, our alliances with convenience stores involve many experiments. Finding partners who will bear some of the risk while promoting the business will be the key to success.

When we were considering forming alliances with convenience stores, while new ideas were pouring in, the number of people attending our meetings also was rising. At the press conference held to announce the creation of the new company, i-Convenience Inc., the presidents of Matsushita Electric Industrial and Mitsubishi Corporation, Kunio Nakamura and Mikio Sasaki, respectively, stood alongside Lawson's Kenji Fujihara to express all three corporations' strong commitment to this project.

Doesn't that sound familiar? Emergence and self-organization have already begun occurring in this complex system.

4.8 Alliances to Strengthen Electronic Settlement

In addition to the technology alliances, portal alliances, and platform alliances described so far, DoCoMo has also built alliances that strengthen our electronic settlement capabilities. In June 2000, we announced that we were investing in a new electronic settlements company, Payment First Corporation.

Payment First is a company that uses electronic settlement via the Internet to manage electronic wallets for its users. The terminals used by its customers are not confined to mobile phones. We anticipate use of its services via computers and home video-game machines as well.

The benefit to participating financial institutions is that Payment First acts as their agent in distributing electronic wallets, responding to queries, and contracting with virtual malls, thus sharply reducing operating costs.

Virtual malls no longer have to sign individual contracts with credit and debit card companies. To take care of settlement issues, financial institutions and virtual malls need only one contract, the one they sign with Payment First.

On the customer side, the use of an electronic wallet reduces the work required to carry out electronic transactions. Until now it has been necessary to install software that conforms to international Secure Electronic Transaction (SET) standards for three-party transactions, but complexity of installation has been an obstacle to the spread of electronic settlements.

By using Payment First, service providers can offer i-mode shopping or other e-commerce services without depending on DoCoMo to handle billing. DoCoMo has invested in this business as a way of promoting the spread of electronic settlements. It is our hope that through this project we can create a more secure marketplace for electronic commerce.

As I wrote at the start of this chapter, we must enlist the help of other companies to implement our strategy of making i-mode the core of everyday life. We will continue to approach other industries and other potential partners, with the aim of continuing to expand our network of alliances.

Chapter 5
Effects

5.1 A Team of Individuals Working with the Same Ideas, Hand-in-Hand, is Stronger Than One Following a Single Boss

As I stated at the beginning of Chapter 1, IT businesses never grow as expected. They are complex systems; once positive feedback begins, the component elements start to exhibit emergence and self-organization, with results that can be far better than would have been predicted.

The converse can also be true: a stumble of some sort can touch off negative feedback and produce unpredictably bad results. Constantly confounding expectations by generating unpredictable results is, indeed, a characteristic of the complex systems model, with its many intertwined, interacting components.

The pace at which i-mode has been attracting new subscribers has, in fact, far outstripped initial projections. When we proposed the basic strategy for i-mode, we were working with a scenario that assumed that each step would occur at certain milestones – when we had one million subscribers, ten million subscribers, and so on. But the emergence and self-organization exhibited by the elements that make up i-mode have gone far beyond our initial expectations. Within each of the communities involved in i-mode – the voluntary sites set up independently, the corporate users, DoCoMo in-house, the mobile phone manufacturers, the media, and the subscribers – events that we had not anticipated have occurred at a pace we could not have begun to imagine.

5.2 The Information Convenience Store and Gemlike Speciality Sites

Let us look first at the rise of the voluntary sites. DoCoMo operates the official sites in cooperation with a host of service providers. The voluntary sites, a separate category, not operated by DoCoMo, have boomed to rival the official sites.

Voluntary sites started appearing as soon as we launched the i-mode service. In our initial scenario, we thought that, to attract one million subscribers, DoCoMo would need to give a forceful push to content development and provide good examples of what the content suitable for i-mode should be. To my surprise, however, some service providers were ready to launch voluntary sites for i-mode from the very start of service, when we had almost no i-mode subscribers at all.

Since then, the number of voluntary i-mode sites has continued to soar (Figure 5.1). Digital Street, which offers an i-mode search engine, reports that as of the end of October 2000, there were about 28 000 voluntary i-mode sites.

At DoCoMo, we think of the difference between the official sites and the voluntary sites as follows. The official sites are like the stock of a

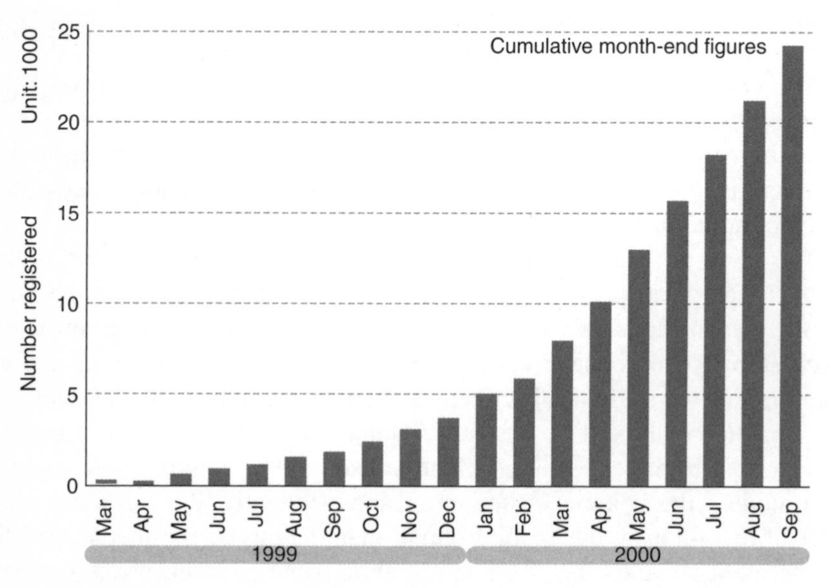

Figure 5.1 Number of voluntary i-mode sites registered. *Source*: Digital Street.

convenience store. DoCoMo is in charge of purchasing for our store and imposes limits on the merchandise to be displayed on its shelves (the service menu items). We require that all items stocked must have mass appeal, must be directed at a relatively large number of subscribers. They must also, of course, not be injurious to good public order and morals.

Because the official sites have to meet our mass audience requirement, they do not include highly specialized gems directed at only a few subscribers. We are running a convenience store, not a boutique-filled, up-market department store.

That is where the voluntary sites come in. One of the virtues of the Internet is that it makes it painless to provide information to only a few people. Our voluntary sites, based on that principle, provide information to limited sets of people, information that DoCoMo's official sites cannot provide.

The number of voluntary sites has grown enormously, just as the array of official sites has. The number of times subscribers access them is also rising. At present, hits on the voluntary sites are running neck and neck with those on the official menu (Figure 5.2). The ratio of total hits on i-mode sites that the voluntary sites attract indicates the depth and breadth of the information available on them.

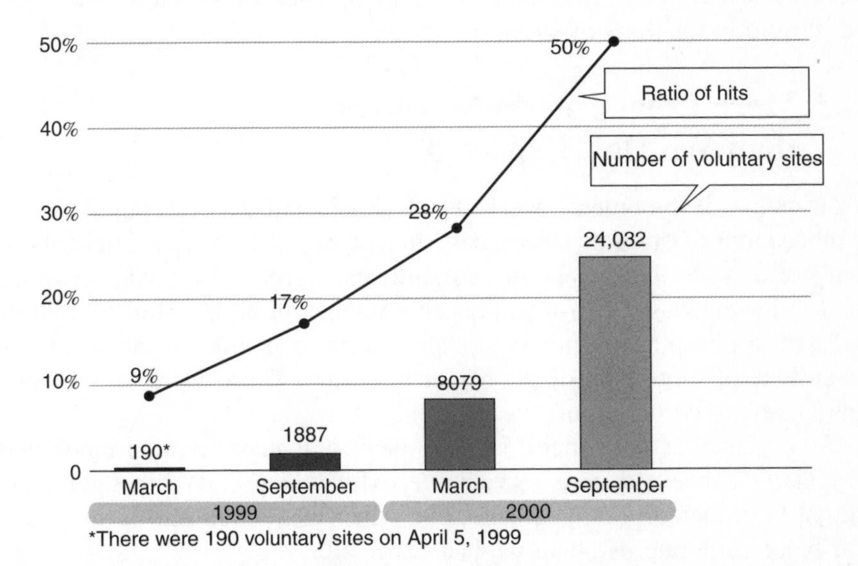

Figure 5.2 Ratio of hits on voluntary sites to total i-mode site hits. *Sources*: DoCoMo and Digital Street.

5.2.1 Eleven Days Later – A Search Engine

Just as surprising as how swiftly the voluntary sites were launched was how quickly a search engine for them became available. A search engine site is built around a program that lets people find the sites they want efficiently by searching for specified keywords and returning a list of the sites where the keywords were found. On the wired Internet, the many search engine sites include Yahoo!, AltaVista, and Infoseek.

It was on February 22, 1999, that we launched i-mode services. Just eleven days later, on March 5, Digital Street launched its Oh! New? search engine service. We had faith that i-mode would become huge and that it would attract search engines just as on the wired Internet, but we had not really expected one to appear quite so quickly.

I learned afterwards that Takateru Imaizumi and his brother Takehiko, who operate Oh! New?, had heard about the launch of i-mode, wrote the search engine together, and then quit their day jobs, such was their confidence in i-mode's success. That is a story that really pleased me.

Oh! New? reminds me of the birth of Yahoo!, the wired Internet search engine. Yahoo! started back in December 1994, when there were few Web sites indeed – and then the Web spread like wildfire. With its first search engine, i-mode seemed to be duplicating that experience. The service had barely started, but as the number of i-mode subscribers grew, we began to receive intimations of success to come.

5.3 More Corporate Subscribers than We Had Expected

One of the developments we had not anticipated was the popularity of i-mode among corporate subscribers. In fact, of our 14 million subscribers, corporate subscribers make up a significant segment. One way to gauge that is by purchases of i-mode-capable phones in bulk. From mid-2000, we saw a conspicuous number of purchases of hundreds, thousands, or even tens of thousands of phones at one time. Those were clearly bulk purchases by corporations.

As a glance at the official i-mode menu will make clear, i-mode was originally conceived as a service for individual use. We thought there might be some corporate demand, but, when we looked into it, we found far more corporate use than we had expected.

For example, Kunio Nakamura, the president of Matsushita Electric Industrial, the consumer electronics giant that manufactures the Panasonic and other brands, uses an i-mode phone. He is an expert at writing e-mail messages on his i-mode phone and uses it to send instructions to other executives. When I met him recently, he joked, 'If you had a contest for inputting e-mail messages on a mobile phone, I'd win in the over-60 age group.' Apparently, after he became president in July 2000, he decided that all senior executives who attend management committee meetings and all division heads should have i-mode phones – about 450 people. The idea was that they could seize the moment, for example, between meetings, to shoot off quick directives. (*Weekend Business* (*Asahi Shimbun*), evening edition, December 16, 2000.)

5.4 Motorcycle Couriers and Exit Interviews

Corporations are using i-mode for more than e-mail. They are using it as terminals for a variety of business applications – internal e-mail forwarding, schedule-management groupware, and order management. For example, the courier service DAT Japan, an early adopter, uses i-mode phones as terminals for its dispatching system. Instructions from the dispatch center to the motorcycle courier, reports on progress in making pickups, and so on go by i-mode. The company had been using both pagers and mobile phones to handle messages between the dispatching center and its couriers, but with the switch to i-mode, they were able to cut their communications costs to a seventh of the earlier figure.

Matsushita and DAT Japan are not exceptional – a very large number of companies are using i-mode in their businesses. Osakaya, a book wholesaler, uses it to allow bookstores to check their inventory and place orders. Tokyo Gas is using it experimentally as a terminal for displaying maps for their service people to check the location of gas pipes when they are out making calls. (The sources of the above examples are 'On the leading edge of the mobile Internet with mobile phones', *Nikkei Communication*, June 19, 2000; 'i-mode, PDAs... how to use them in business', *Nikkei Open System*, June 2000; and 'Business effects of i-mode use', *Nikkei Information Strategy*, September 2000.)

i-mode phones were also used in conducting exit interviews at polling places in the August 2000 elections for the lower house of the Diet.

While we have information on business uses of i-mode through newspaper and magazine articles as well as direct comments from users, DoCoMo actually has no precise grasp of the extent to which companies are using i-mode. Many corporations have built systems using i-mode with no direct involvement on DoCoMo's part. That in itself is a triumph for our vision: because we chose to use the standard markup language, HyperText Markup Language (HTML), they could build their systems on their own quite easily, without asking DoCoMo for help.

5.5 Self-Organization in Off-the-Shelf Software

Still, there was a phase when, as DoCoMo's Gateway Business division, we were actively working to wangle business use of i-mode, before its launch. We made approaches to several companies developing off-the-shelf software for business use, asking what they thought about our ideas for using i-mode. We were hoping to get one to take off as the lead goose, as it were, to go back to our complex systems model.

The first to respond was Pumatech, Inc., a software company with its headquarters in the United States. On February 1, just before i-mode services began, Pumatech and DoCoMo announced they were developing Intellisync Anywhere server software for corporate use. Installing that software makes it possible to check e-mail messages and meeting schedules on Microsoft Exchange or Lotus Notes and Domino from an i-mode phone instead of a personal computer.

Once Pumatech had started the ball rolling with a gateway server, Intellisync Anywhere, producers of groupware began to respond – Lotus Notes and Domino, NEC's StarOffice, and Compaq's Bizport. Today almost all groupware packages have some degree of i-mode support (Figure 5.3).

That, again, was an instance of emergence and self-organization in a complex system. Initially, using i-mode, which was designed for consumer use, companies came up with the emergent concept that they could apply it to their own businesses. And companies that learned through the media that other companies were using it wondered if they could introduce it as well and so, acting independently but with one accord, they began building systems – and we had self-organization in action.

5.6 Another Effect of Adopting HTML

Companies introducing business systems using i-mode were not doing so out of concern for DoCoMo's bottom line. They were interested

Company	Product	Date shipments began
NEC	StarOffice/PocketWeb for i-mode	October 1999
Big Bang System Corporation	ExLook for exchange	May 2000
Hitachi	Groupmax Mobile Option for i-mode	April 2000
Fujitsu	TeamWARE Office via iMode Option	January 2000
Pumatech	Intellisync Anywhere for Microsoft Exchange	March 2000
Lotus	Domino PhoneConnect	December 1999
NTT data	Intra-Mart Intranet Start Pack	October 1999
Compaq	Bizport	August 1999
10art-ni	la couleur/rouge2	August 1999
DreamArts	Insuite99	September 1999
Neo Japan	iOffice2000	July 1999
Orangesoft	iModeGate	June 2000

Figure 5.3 Major groupware and e-mail client software supporting i-mode. *Source*: 'On the leading edge of the mobile Internet with mobile phones', *Nikkei Communication*, June 19, 2000, pp. 118–19.

in making their own systems more flexible and easier to use and in reducing system-operating costs. The result was, however, advantageous for DoCoMo in that it increased our packet communications revenues. Here, too, the complex systems model works.

In observing those examples of emergence and self-organization, I appreciate how important our decision to use HTML, a nonproprietary markup language, was. I can imagine the systems people at corporations and developers of packaged software, when asked by their customers or superiors, 'Can't this software be used from an i-mode terminal?', replying, 'Sure. Because i-mode's markup language is based on HTML, that would be a snap.' And then, if they were told to go for it, it would be as good as done. For example, it took less than a month for DAT Japan, the courier service I mentioned earlier, to put together its i-mode-based dispatching system. Obviously, it depends on the scale of the system, but some companies have said that they developed their systems in a matter of days. That, clearly, is the virtue of adopting *de facto* standard technology. What would have happened if we had adopted an independent standard developed for the mobile communications industry? Not the emergence and self-organization we have observed.

5.7 A Fast Conversion from a Telecom to an Internet Company

The launch of i-mode touched off major changes at DoCoMo as well. The biggest internal effect was that, through i-mode, our understanding of what it is to be an Internet business advanced far more than we had expected. DoCoMo was, of course, originally a telecom. But upon starting i-mode, we have been transforming ourselves into an Internet company – both in the speed with which we do things and in our thinking.

Consider, for example, the nationwide meetings we instituted to vet content for the official sites, as I described in Chapter 3. Our Gateway Business division is in charge of i-mode, but initially only a very few of our personnel understood the essence or special characteristics of the Internet or had any understanding of what content ought to be available on the official i-mode menu. Those members who did understand led the discussion of candidates for the official menu at the nationwide meetings that decided which were suitable and which needed revisions and improvements. While some thought the meetings were wasteful, I think that the repeated discussions at these meetings produced a significant effect. That is, the meetings were not to decide whether the proposed content was good or bad. The discussion was framed in terms of whether it would satisfy our subscribers or not. One positive by-product of that process was that more people learned how to evaluate content in terms of the nature of i-mode as a service and what DoCoMo should consider concerning content. Instead of a few people in the Gateway Business division, we had content managers throughout the country sharing that new perception of the characteristics of the Internet as a medium.

Lately I have heard, 'You know, Mr X of the regional company is saying the same kind of things you do.' And when I ask what Mr X was saying, it is often much more pointed than what I myself might have said – but definitely on the same track. That is how far we have come in developing shared consciousness within our company.

The regional content managers who have participated in our nationwide meetings presumably explain to their fellow employees what the nature of i-mode as a service is and what an Internet way of thinking is. Thus, more and more people inside the company share the same ideas – the complex systems model in action.

5.8 An Organization of Empowered Individuals is a More Powerful Organization

The role of the nationwide meetings then was to build a shared way of thinking about what sort of service i-mode is. We left the problem of deciding how to act on the basis of that thinking up to the various managers. To give its components a degree of freedom and yet be able to move an organization is also one of the characteristics of the complex systems model. Each individual acts according to his own understanding of what i-mode can be, adds his or her own tweaks and innovative touches; as those actions lead to results, all participants feel they have personally contributed to the success of i-mode.

One aspect of those results in i-mode can be seen in the marketing activities. In DoCoMo's case, we not only have the central company, whose service area is the seven prefectures of the greater Tokyo metropolitan area plus Yamanashi, Nagano, and Niigata prefectures, but also our regional companies, each of which has its own planning and advertising and PR departments. Each carries out its own i-mode marketing activities. The central company briefs the relevant personnel in the regional companies on the nature of the i-mode service and our basic policies but does not give detailed instructions on how to market it.

Thus, the people in charge of marketing at each of the regional companies think it through and then try out various ideas. The result is that each of our regional companies creates unique advertising, which helps attract more subscribers. For example, both DoCoMo Kansai and DoCoMo Kyushu chose to hire celebrities with strong regional ties (Hanami Honjo and Reina Tanaka) to appear in their i-mode commercials – with good effect.

In fact, it was the Kansai, in Western Japan, where i-mode first took off. Then the excitement spread nationwide and led to i-mode's currently thriving state. Actually, the mobile phone wars were fiercer in the Kansai than in the Tokyo area. If all the marketing there, where the struggle for mobile phone market share was so difficult, had been guided by the central company's approaches, where would we be today? I am inclined to attribute i-mode's swift takeoff there to inspired marketing by DoCoMo Kansai.

As that example suggests, I believe it is much more powerful for an organization to have many people who share the same attitudes and ideas than to have one individual with superb leadership qualities and abilities act arbitrarily in leading the whole organization. I am not sure what proportion of the members of an organization need to share the same thinking for this model to work – 30%, 50%? – but when it does, an organization of empowered individuals is far more powerful.

There lies the critical difference in getting an organization to act. Structuring the situation so that each individual spontaneously participates generates almost unimaginable power.

5.9 Accelerated Mobile Phone Evolution

The pace at which mobile phones evolved also far exceeded our predictions. As I said in Chapter 3, technology tends to converge on products that can sell in large quantities. Because there are simply more mobile phones than personal computers, I expected that the speed at which personal computers were miniaturized would be slower than the speed at which mobile phones gained advanced features. The results proved me right. Since the launch of i-mode, the pace of development of new mobile phones has been picking up. Because manufacturers have realized that i-mode phones sell, they are pouring substantial resources – both human and monetary – into their mobile phone development divisions.

The results can be seen in the number of mobile phone models available and their development cycle. During the first year of i-mode service, there were only four i-mode-capable mobile phone models, the 501i series produced by NEC, Fujitsu, Matsushita Communication Industrial, and Mitsubishi Electric. Those companies were also producing other mobile phone models during the development phase of their i-mode models and were quite reluctant to allocate resources for developing the i-mode phones.

But when the number of i-mode subscribers started to grow, resources started converging on the i-mode models. More manufacturers offered their versions of the second generation i-mode phone, the 502i, and the basic series of mobile phones, starting with the 209i when it went on sale in June 2000, were also made i-mode-capable. The 82li Super Doccimo series, which could be used as a Personal Handyphone System (PHS) phone, also joined the i-mode line (Figure 5.4).

As of October 2000, a total of 21 i-mode models have been launched. Today it seems unbelievable that it took so long for the manufacturers to understand the i-mode concept as we explained it, when we were asking them to develop the 501i series.

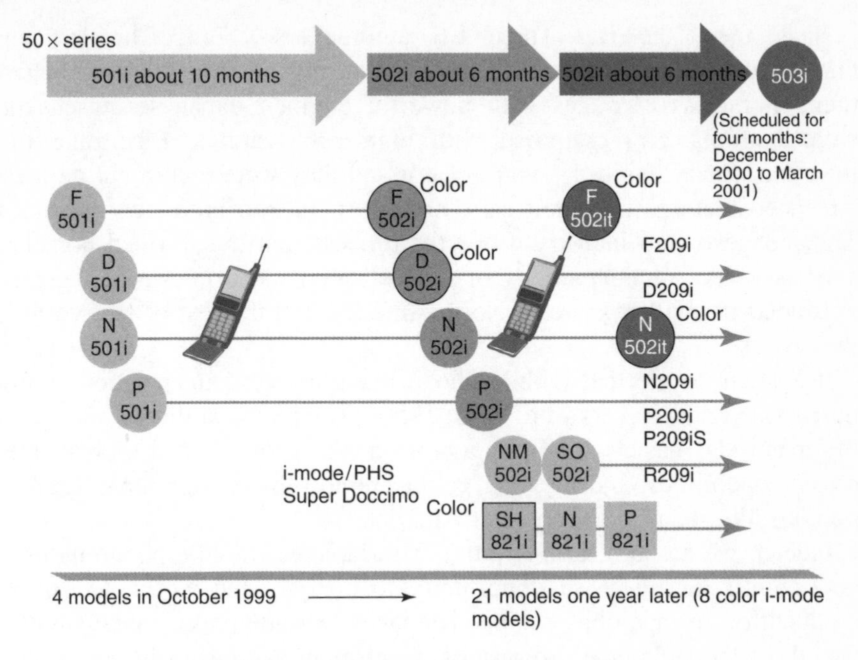

Figure 5.4 The mobile phone development cycle shortens.

With such a variety of models on sale, each manufacturer must come up with distinctive features of its own in order for their models to sell. Some have chosen the tack of significantly reducing the weight of their phones, others have focused on innovative design, and others have tried adding value with, for example, built-in music playback features. And, of course, their search for differentiation goes on.

5.10 Like the Auto Industry in the 1980s

The result of that competition is the dazzling technical progress that mobile phone manufacturers – and their sources of components – are effecting. Mobile phones are growing smaller and lighter by leaps and bounds, and the trend to more advanced features – including color displays and the ability to play back music with harmony – is accelerating.

I am constantly telling Japanese mobile phone manufacturers that this is their big chance. When I look at the mobile phone industry now, I feel I am seeing a replay of what happened when Toyota and other Japanese auto manufacturers made their move into the American market in the 1980s.

Back then, the Big Three US automakers – Ford, Chrysler, and GM – were making cars that were, in terms of the mind-set behind them, horseless carriages with powerful engines. Japanese automakers were building cars crammed with high-tech features. Electronic fuel injection, rotary engines – you name it and they were incorporating these advanced features into their cars, turning them into high-tech machines. The Japanese auto industry drove the process that transformed horseless carriages into amalgamations of sophisticated electronics – and greatly expanded its market share in North America and the rest of the world in the process.

It appears to me that mobile phone manufacturers are positioned now much as the automakers were in the 1980s. The plain vanilla mobile phone in Japan today has a color LCD, a built-in Web browser, and sophisticated music capabilities, and they are getting rapidly more high tech. There is nothing like them in Europe and America.

Indeed, we are at a turning point. The Japanese mobile phone industry is, I think, staring at an opportunity to conquer the world. Cut-throat competition in Japan has equipped Japanese mobile phone makers with a wealth of technological prowess of which they can be justly proud, new technological innovations are occurring, and I expect them to make great strides in the global market.

5.11 The Media Rush

Another development that exceeded our initial expectations was the media excitement over i-mode. As i-mode began to attract more subscribers, the number of books and other publications about it grew. According to a study by Nippon Shuppan Hanbai, Inc., a major book and magazine distributor and retailer, as of the end of October 2000, 102 books had been published about i-mode (Figure 5.5). Because magazines have been bringing out extra issues dedicated to i-mode as well, it is difficult for us at DoCoMo to keep track of them all. Beyond the books and the magazines that we might expect to feature coverage in this field – the trend-spotting and computer magazines – general-interest men's magazines and women's magazines have been churning out special features on i-mode as well.

These books and special issues cover a range of i-mode topics: how to choose an i-mode phone, what the voluntary sites are, how to build a Web

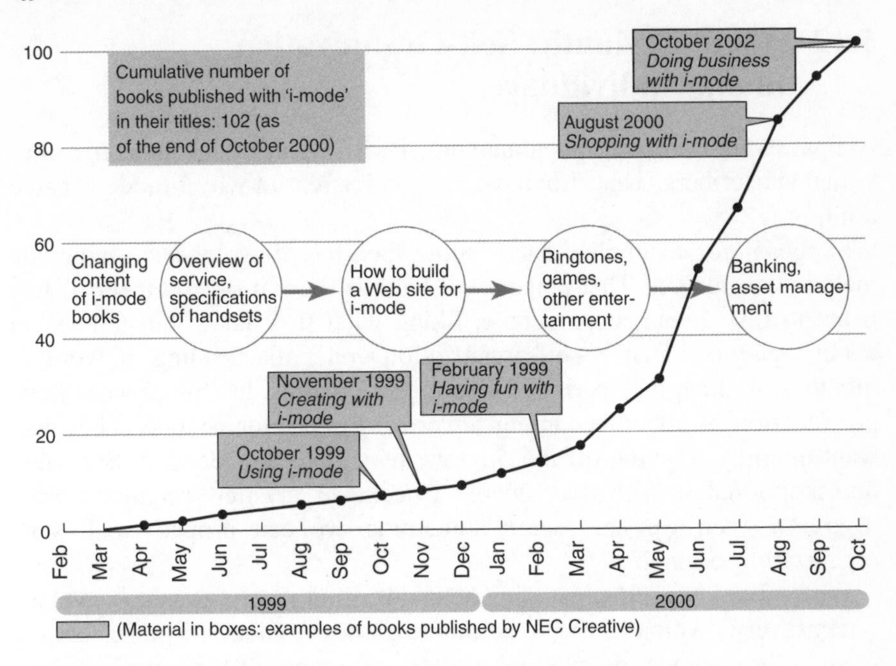

Figure 5.5 Books on i-mode. *Source*: Nippon Shuppan Hanbai, Inc.

site for i-mode – you name it. In many cases, DoCoMo was approached by the writer and cooperated in providing information, but lately books and magazine articles about i-mode have been popping up without bothering us at all. We, of course, very much welcome this self-starting behavior on the part of the media. Having the media jump on the i-mode story has raised people's awareness and interest and helps boost our number of subscribers.

Quite apart from books and magazines, it seems that hardly a day goes by without a newspaper article on i-mode. 'New i-mode service' is the lead, whether the service is on the official menu or is one of the many voluntary sites. The newspapers are also full of stories on 'How we use i-mode in our in-house system', or 'Tie-up over content technology for i-mode', and other business articles. That, I am quite sure, has led top executives in other companies to wonder if they should be doing something with i-mode too. In fact, some corporations that stayed away from e-commerce on the wired Internet have begun doing business on i-mode. Here, too, we are seeing self-organization in process, pulling in the media and then having wider effects.

5.12 Word of Mouth: Self-Organization among Individuals

A final example of the phenomenon of self-organization concerns individual subscribers. That, I believe, is a major reason why i-mode became a hit.

Suppose, for example, that a subscriber has downloaded a ringtone melody and likes it. That person will tend to show it off to others – 'Just listen to this' – and those people, liking what they hear, will tell others about that tune. That is self-organization under the heading of word of mouth – or bragging, perhaps. Those participating in this process have no intention at all of becoming advertising flaks for i-mode. They just want to enjoy showing off the ringtone they have downloaded. But when one individual says to five others, 'Check out my new ringtone – isn't it great?' then we are seeing something we can properly call self-organization occurring.

What stimulates self-organizing behavior is not just ringtones; it can be a screensaver download service, an online game, or another i-mode service. Even before i-mode, people took pride in showing off the mobile phones they carried, and as the phones have become increasingly multifunctional, that tendency seems to have intensified.

Why has the number of i-mode subscribers mushroomed to 14 million, with a very high proportion of those subscribers using the services available on it very often? The answer, I believe, is that we are seeing the result of self-organization on the part of subscribers.

Chapter 6
The Future

6.1 Emergence and Self-Organization on a Global Scale Give Rise to Greater Positive Feedback

Three years have passed – in a flash – since I became involved in the i-mode project. During this period, we defined our objective as making i-mode the core of everyday life, and took steps to make that happen. Our cooperative relationships with many other corporations in many fields have helped us approach our objective, but we are not quite there yet. We must not mark time but move forward, widening the span of our partnerships.

The next-generation i-mode strategies we are now considering include the roll out of third-generation (3G) i-mode service using the IMT-2000 standard, developing e-commerce, including e-Money, and implementing an international strategy.

In terms of the categories of progress outlined in Chapter 4, 3G i-mode service will be an extension of the technology alliances that have followed the development of more sophisticated mobile phones, while e-commerce and going international will be an extension of our platform alliances. Let us look at the strategies and directions involved in each case.

6.2 The IMT-2000 Age is Drawing Near

DoCoMo plans to launch 3G service in May 2001, starting service in the 23 wards of Tokyo and in the neighboring cities of Yokohama and

Kawasaki, using the IMT-2000 standard, also known as W-CDMA, which stands for wideband Code Division Multiple Access (CDMA), a high-speed 3G mobile wireless technology. We will gradually expand the 3G service area so that by April 2002, all our regional companies will be starting to offer it, initially in the area around the capitals of each prefecture. We will then expand coverage in each region to achieve coverage of 80% of Japan's population by March 2004 (Figure 6.1).

The new service has two main distinctive features. It supports faster transmission speeds, from 65 to 384 kbps, and it will permit using the same mobile phone in Japan and overseas. The latter is what is called *international roaming*.

At present, i-mode operates at a transmission speed of 9.6 kbps. The 3G service will accelerate transmission by up to 40 times. That will be fast enough for streaming video service – not possible on existing mobile phones – and for videophones as well. We have decided to adopt MPEG-4 (Moving Picture Experts Group-4) technology – the international standard – for our data-coding standard for streaming video images.

In terms of types of handsets, we expect to offer a standard i-mode phone that is an extension of the handsets currently in use, with the focus on voice functions, plus a visual phone designed particularly for sending and receiving image data, and a dedicated wireless data terminal, in the PC card format, that can be inserted into the PC card slot of a notebook computer. Each will support asynchronous uplink/downlink packet communications speeds. The maximum downlink speed (network to subscriber) is 384 kbps; the uplink maximum (subscriber to network) is 64 kbps. The standard phone will use packet communications for all data transmissions, while the visual phone will use both packet communications and direct connections.

Date	Service
May 2001	Started in 23 Tokyo wards, Yokohama, Kawasaki
December 2001	Started by NTT DoCoMo Tokai and NTT DoCoMo Kansai
April 2002	Started by NTT DoCoMo Hokkaido, NTT DoCoMo Tohoku, NTT DoCoMo Hokuriku, NTT DoCoMo Chugoku, NTT DoCoMo Shikoku, NTT DoCoMo Kyushu
March 2004	Service available for 80% of the population

Figure 6.1 The schedule for rolling out W-CDMA (IMT-2000) G3 service. *Source*: DoCoMo.

We are planning to implement 3G in two stages, in May and November 2001. First, in May, we will offer W-CDMA (IMT-2000) services in part of the Tokyo metropolitan area. Step one will be the launch of sales of i-mode phones that can support the faster transmission speeds. At that point, the advantage to subscribers will be that they can download the same content as in the past, but faster. Also, because the new phones will have larger memories, service providers can develop more complex programs for downloading. Then, in November 2001, we plan to roll out service not only in the Tokyo area but also in the two other major metropolitan areas, Nagoya and Osaka. At that point, sales of i-mode phones that have software built in to play back MPEG-4 video will begin. Subscribers using the new i-mode terminals will be able to enjoy content with video streaming.

Furthermore, 3G supports international roaming. DoCoMo has encouraged the adoption of the W-CDMA wireless technology, as have European telecommunications providers and manufacturers. If telecoms overseas launch the same type of service, then it will be possible to use the same mobile phone at home and while traveling abroad. Until now, each country has adopted a different wireless standard, and thus it was not possible to use one's mobile phone in other countries.

Both high-speed transmission and international roaming will significantly increase convenience for our subscribers.

6.3 3G Marketing: 'First, the Applications'

All 3G terminals, except the data card for use with notebook PCs, will be i-mode-capable. That is, i-mode will be standard on all mobile phones intended for human use. From a technical perspective, these mobiles can be described as offering the next generation in mobile communications services, with high-speed transmission (64 kbps and higher) and MPEG-4 compression of moving images. But we will not deviate from our i-mode strategy, which is to say, we will not be focusing on the technology.

Instead, the message to beguile customers will be, 'the high-speed i-mode with video clip feature'. That is, 3G is i-mode with the ability to let you see short clips of a full motion video. This we will present as an extension of existing i-mode services.

Most of our subscribers are utterly uninterested in what makes i-mode possible, whether it is the personal digital cellular (PDC) format now in use or a new 3G format. They also are unlikely to have much sense of what it means to say that the transmission speed will increase from the current 9.6 to 64 and 384 kbps.

What subscribers do care about is applications. Applications are what matter. The decider is not what technologies are used but what the subscriber can do with those technologies. That is why video clips are going to be our core message about G3.

The way we envision the service is that there will be a video icon on the i-mode screen. Press it, and video streaming starts. Most of these video clips will be from a few seconds to a few dozen seconds long.

Many people will assume that with a service transmitting at 64 kbps and above, they can download videophone images, music, or full-motion video clips lasting minutes or more. But no matter how much faster video images can be transmitted over a wireless network, it is never going to match, in terms of cost and image quality, what you can see on your television set free of charge. While it is true that wireless, mobile communications eliminate constraints on time and place, how many subscribers will want to spend the time and money to use the new services? Will the content be good enough to attract them? Even more than now, when text data is the main kind of content being provided, we probably will find that striking a balance between the services we offer and the cost to the subscriber will be critical.

6.4 Seconds of Real Enjoyment

I believe that, even working under tight time constraints, there are many kinds of content that can be enjoyed thoroughly in video clips lasting well under a minute. If the clips are kept to that length, the packet communications charge would be no more than ¥100. This is inexpensive enough that a relatively large number of subscribers would download them.

Consider, for example, sports highlights – the home run, the goal being scored. Those scenes would be excellent content for i-mode video clips. Recently, with my eyes peeled for new services, I have been clocking times when watching the sports news. From the pitcher's windup to the bat connecting to the ball flying into the stands lasts, at most, about 10 seconds. A goal in soccer happens about as quickly.

Video clips running a half-minute or so might include promotions for new movies or newly available videotapes, interviews with celebrities, commercials, and other content. Even a clip well under a minute would be quite satisfying.

Those are just ideas for services using video clips – but DoCoMo is not exactly a professional in the video content field, and our ideas are limited. What we hope is that large numbers of service providers will join us to develop video applications that we would never have predicted.

Out there in, for example, television stations around the world are professionals in the editing of video images so that they are effective in a brief period of time. One of the key skills in television broadcasting is editing a long recording so that images can be broadcast efficiently in short clips of a few seconds or so, without losing impact. The clips shown with the news, not to mention television commercials, are all very short, ranging from a few seconds to half a minute. In fact, Japanese broadcasters already have i-mode sites for their news programs. We anticipate that they will quickly begin to offer video-streaming services as well.

Of course, DoCoMo will meet the demand not only for downloading streaming video but also for interactive applications, such as videophones. As a high-end handset assumed to be used for transmitting several minutes or more of video, we will offer our visual phone, which will be a separate series from the standard model. The visual phone will use direct connections, which will be less expensive than packet communications for longer transmissions.

6.5 Striking a Balance between Subscribers and Service Providers

As I have said, subscribers will not go along with dramatic changes made in a single bound. That is a point to be borne in mind in creating the content for G3: even if it is possible to transmit streaming video, subscribers will not necessarily leap to have that feature. The existing i-mode service offers text-based sports updates and news programs. If the video icon were interpolated between segments of those news transmissions, then subscribers would be readily inclined to accept the innovation. The starting point should be that subscribers who are particularly interested in a news item they are reading can click on the icon and see the video as well. Presenting a combination of text and video will be, I think, the key to making the new service appealing.

The wired Internet already offers services that combine text with video clips. CNN, for example, streams video clips of the news on its Website. DoCoMo will do well to consider how these and similar services are built and presented.

Apart from not making the new service too alien for subscribers, we need to keep the barriers to entry low for content providers too. That, fortunately, should not be a problem, since those already providing streaming video in the wired Internet need only prepare content for i-mode on the same servers. We are looking forward to seeing what content providers' creativity will lead to in full-motion video offerings.

What is critical, in providing this new service, is that DoCoMo, the content providers, and the subscribers move ahead at the same speed and in good balance. If one gets too far ahead of the other two, the service will not take hold. It is up to DoCoMo to coordinate skillfully so that video-streaming services are well received by subscribers.

6.6 e-Money from Mobile Phones

The second of our next-generation i-mode strategies is aimed at making e-Money a reality. As I explained in Chapter 3, the wallet PC was one of the defining images for the i-mode handset. This implies that, ultimately, we are aiming at implementing e-Money on i-mode. That is something many are working on in the wired Internet, for its virtual malls and other e-commerce sites. Actually, though, for the big-ticket items that virtual malls stock, payment by credit card works well. The buyer can input the card number, payment is on the next month's bill rather than instantly, and in some circumstances a cooling-off period applies. That is very convenient for the purchaser and imposes little risk on the mall operator. Thus, there does not seem much opportunity for introducing e-Money in virtual mall operations.

Where e-Money is attractive is in simplifying small purchases, the range of purchases for which accepting payment by credit card does not make sense for the seller, given the fees that credit card companies charge. That is where e-Money can work.

I think that the first full-scale introduction of e-Money will come in points of physical contact between the mobile phone and the seller: mobile phones and convenience stores or mobile phones and vending machines, for example.

Many people do not like having to haul out change for every small purchase – especially on, for example, their lunch breaks, when the lines at convenience store checkouts are long and slow. The solution is to reduce the number of situations in which people have to use change. Since convenience stores and vending machines are where people often use change, enabling payment by e-Money will free people from much of the tedium that dealing with small change entails.

For the convenience stores, having to handle less change would mean more efficient use of their POS-equipped registers and being able to serve more customers. The task of counting up each day's sales would also be simplified. Thus, introducing e-Money operating between a mobile phone and a convenience store would make both the store and its customers happy. That is why I think e-Money will take off in convenience stores.

6.6.1 The Role of Java in Popularizing e-Money

At this point, however, there are no signs that the many companies developing e-Money formats will agree on a unified format. It is going to take considerable time before the convenience store chains and other service providers agree on a standard for e-Money so that it can be used everywhere.

To facilitate development of the *de facto* standard, it is conceivable that DoCoMo itself could issue a virtual currency, 'DoCoMo Money', and call on the various service providers to adopt it. But, speaking for myself, I think that is not very practicable, because that would be well out of DoCoMo's areas of expertise, it would take us a long time to develop an e-Money product that would be really convenient, and, of course, there would be no guarantee that retailers and other service providers would adopt our standard, once we had developed e-Money to our satisfaction. For DoCoMo, it is more realistic to concentrate on providing a platform that can handle the various kinds of e-Money providers decide to issue.

Having Java-capable browsers will facilitate the spread of e-Money. Using Java, it will be possible to distribute software for handling the e-Money each chain issues in the form of Java applets, one for each chain. The Lawson chain of convenience stores, for example, would be able to accept Lawson Money with Lawson Wallet Software issued to mobile phones. The same mobiles could also handle e-Money currencies issued by other chains.

Java alone will not be sufficient to popularize e-Money. Having started thinking in terms of mobile phones handling e-Money, what we are now focusing on is the external interface – the short-range wireless interface connecting the mobile phone and a cash register or vending machine. The candidates for that wireless interface include the Infrared Data Association (IrDA) standard widely used in notebook computers and personal digital assistants (PDAs), Bluetooth, a new wireless standard now beginning to

Interface	IrDA	Bluetooth	Contactless IC card (ISO/IEC14443)
Frequency	Infrared	2.4 GHz	13.56 MHz
Standards organization	Infrared Data Association	Bluetooth SIG	ISO, IEC
Transmission speed	4 Mbit/second (Ir DA 1.1)	1 Mbit/second	106 kbit/second and above
Range	3.8 m	10 m (at 1 mW)	10 cm or less
Examples of uses	Personal computers, PDAs	Between personal computers and printers and between mobile phones and handsets	Telephone cards for pay phones, public transit systems

Figure 6.2 Wireless short-range interfaces.

be widely adopted, and the contactless IC card approach that the JR and other railways are considering for adoption at ticket offices (Figure 6.2).

If the wireless interface to connect mobile phones to convenience store cash registers, for example, can be ironed out, then it will be possible to handle e-Money transactions using that interface. And because a three-way relationship can be set up connecting the mobile phone, the register, and the network, the transaction can be confirmed on request from either or both the mobile phone and the cash register.

Some argue that introducing local wireless connections for using e-Money would reduce traffic over DoCoMo, but I doubt it. The more opportunities our subscribers have to use their i-mode phones, the more the traffic grows. Actually, the local wireless connection is essential for popularizing e-Money. If paying with e-Money required making a connection to a central point – and paying the resulting communications charges – then subscribers would be reluctant to use e-Money, and traffic would fall, not rise.

6.6.1.1 The Wireless Interface

The three candidates for the short-range wireless interface are, then, IrDA, Bluetooth, and the contactless IC card interface. DoCoMo has not decided which to adopt because at this point there is no consensus on which wireless interface will be the standard equipment in the POS registers and vending machines that our mobile phones would be talking to. One reason no clear direction has emerged is that work on developing the Bluetooth standard, which is very attractive in terms of security and other factors, is not moving as quickly as I had hoped. At this point, the Bluetooth effort is focused on standardization for personal computers and peripherals, and developing specifications for e-Money applications has a low priority.

We should follow the same complex systems logic that we had in the past to provide the standard for selecting an interface. That is, we should consider the technical aspects of IrDA and Bluetooth carefully, but not use technology alone as the basis for decision-making. It is, I think, important to focus instead on what the interface between a vending machine or POS register and a mobile phone should be.

We should also, I think, be making our preparations so that once the direction for the wireless interface for POS systems or vending machines is set, we are ready to move. Because of our alliance with Lawson, for example, we know that transactions using e-Money via a wireless interface linked to the in-store multimedia kiosk or POS register are under consideration. Once the basic direction is set, I want us to be ready to pursue this opportunity aggressively.

At this point, there are no precedents to the use of e-Money in this way. That puts us in an unfamiliar position. Thus far, we have used slight modifications of standards with considerable track records – HTML and MIDI, for example – for i-mode content, but with e-Money, we must start from scratch. That is hard. I hope that with the cooperation of Lawson and other partners, we can make it work.

Before e-Money gains credence, DoCoMo and Lawson are considering ways to utilize 'electronic value'. One possibility that looks likely to be well received would work in the same way as an airline mileage program: the store would give points, electronically, for each purchase. The points would be stored in the memory of the customer's mobile phone and the customer could use them for a discount on a subsequent purchase, on the basis of cumulative points.

The points would give discounts on the order of some tens of yen – sums small enough such that this system would not need the level of security a full e-Money system would. That means lower barriers to participation, for both the stores and those who build the systems to make it work. But discounts of ¥50, for example, would be of considerable value to people who frequently shop at convenience stores.

The benefits of e-Money will be large, but I doubt that consumers will leap to adopt it. Thus we advocate an incremental approach, perhaps starting with discount points, to accustom people to using their mobile phones in making purchases.

6.7 Alliances with Home Appliance Manufacturers: Another Possibility

As another approach to making mobile phones the core of everyday life, we are also considering alliances with manufacturers of home appliances and personal computer peripherals, such as printers. Imagine, for example, an alliance with a VCR manufacturer. It would be possible to find the program you want to record through an i-mode search, retrieve its G-Code for timer programming (G-Codes are like VCR Plus Codes in the United States), and send that code to the VCR, via the wireless interface built into the mobile phone, so that the VCR is set up automatically to record the program.

Applications like that may not happen any time soon, given the replacement cycle for home appliances. Most home appliances are replaced only every 5 or 10 years – and thus we cannot expect to build a trend to appliances interfacing with i-mode phones in a short period of time.

Which interface we choose for the wireless connection is important here as well. Apart from home appliances, the mobile phone could be communicating with POS registers in convenience stores, with vending machines, and other sales outlets. We must consider how our phones will work with all these devices and not make a decision arbitrarily. Our choice should always be the one that is likely to provide our subscribers with maximum convenience.

As I explained in Chapter 4, once alliances with car navigation systems, game machines, or convenience stores start rolling, the number of situations in daily life in which one uses an i-mode phone will multiply. If we can turn all those possibilities into realities, then we will be nearing our goal of making i-mode the core of everyday life. We are already halfway to putting the wallet PC that Bill Gates was predicting four years ago in people's pockets. It should not take too much longer to take it the rest of the way.

6.8 For Export: Our Business Model

Finally, I would like to say a few words about the international expansion of i-mode. Having participated in negotiations for overseas alliances and also given my fair share of lectures abroad about i-mode, I sense that telecommunications providers around the world are keenly interested in starting up a service like i-mode.

All telecommunications providers share the same problem wherever they operate: they are searching for the next scenario, the one that will enable their companies to continue to grow once mobile phone service has spread to the point that the number of subscribers has reached its maximum. When visiting industry colleagues in other countries, I often hear, 'We'd like to introduce a nonvoice service like i-mode. What should we do?'

My response, first of all, is to say, 'We can't take the i-mode networks DoCoMo employs overseas, and even if we could, they wouldn't work well.' I say that for several reasons. One is that Japan and other countries have different mobile communications systems. DoCoMo uses PDC, a variant of Time Division Multiple Access (TDMA) technology used only in Japan. Global System for Mobile Communications (GSM) is used in Europe and Asia, while CDMA and another version of TDMA are used in North America.

Far more important than differences in standards, however, is that i-mode is a business model, not the name of a system. I would not be sanguine about the outcome if the i-mode system were merely transplanted to another network standard, but translating the business model is another question. The ideal outcome would be for our colleagues in other countries to develop a deep understanding of the nature of the i-mode business model, creating an environment in which emergence and self-organization can take place not only among telecommunications providers but also among service providers and manufacturers.

In terms of the timing of i-mode's overseas expansion, I would like to work toward the point at which other countries are starting up their own 3G mobile communications services. Specifically, we will see the first overseas expansion of i-mode in Hong Kong, home base of the Hutchison

Telephone Company Limited, in which DoCoMo has invested, and in the Netherlands, Germany, and Belgium, served by KPN Mobile N.V. Launching i-mode in those markets even before the start of 3G service is also being considered.

6.8.1 Speaking with Greater Authority

DoCoMo would like to promote the overseas adoption of the W-CDMA format we have adopted, but not so that we can sell W-CDMA phones there. We are not in the mobile phone manufacturing business; no matter how much W-CDMA spreads overseas, we will not dominate other markets. Rather, the advantages for DoCoMo of flying the flag for W-CDMA in moving overseas are twofold. One is direct profit. If the other telecommunications providers in which we have invested make a success of the new service, their companies will be worth more, and the value of our investments in them will rise. If we were to sell our share holdings in them, DoCoMo would make a profit.

DoCoMo also wants to continue to fly the flag for W-CDMA to increase our influence in mobile communications by expanding the range of regions and subscribers we serve and thus maintain our influence and authority in the industry. That strategy is similar to Vodafone's. Through the telecoms in which we have invested, the number of subscribers over which we have influence will grow. That in turn produces advantages in negotiating with other potential partners.

The more subscribers DoCoMo serves, the more the handset manufacturers and the providers of new services will listen to DoCoMo. We expect that a series of new services will be launched, on a global scale, in the near future. If DoCoMo is still confined to domestic operations when that happens, then whether we can retain our position as one of the industry's leading companies, worldwide, will be an open question. Competition is relative; there is no such thing as absolute superiority. Because we cannot ignore the possibility of another company that is even bigger and more advanced coming on the scene, DoCoMo must always stay at the leading edge.

At present, DoCoMo is in the group of companies in the vanguard of mobile communications. Thanks to our leadership position, we are approached by an extremely large number of manufacturers and service providers. I want to retain that position. To do so, we must move overseas, to acquire global influence.

6.8.2 First, International Roaming

What, though, is the advantage to our subscribers of DoCoMo's international thrust? Some have asserted that there is nothing in it for domestic subscribers if DoCoMo moves into operating overseas. I disagree. A world in which mobile phone subscribers overseas also use i-mode will be advantageous for our domestic subscribers in many ways.

Before explaining those advantages, I would note that it is because of i-mode that we will be able to pull off our early launch of 3G service in Japan. DoCoMo had always planned to introduce 3G service promptly after the standard was finalized, but thanks to the strong subscriber response to i-mode, the pace at which we are rolling out 3G has picked up. That is because added-value services such as the video clip service I described earlier can be presented as extensions of existing i-mode services.

By contrast, plans for introducing 3G service in other countries, notably in Europe, are proceeding more slowly than in Japan. Even in northern Europe, where the proportion of the population using mobile phones is even higher than in Japan, the schedule for transitioning to 3G is slower.

One reason for that delay is that the European telecoms have yet to sense sufficient demand for data communications services. While they have high expectations, it appears to me that they are not as confident about 3G as DoCoMo is. Manufacturers of handsets, who are counting on 3G for a new source of demand, are pushing for the introduction of 3G service, but it is the telecoms that will ultimately have to take the risk. If they are not fully committed to 3G, then there is no way that putting the new telecommunications infrastructure for it in place – an effort that will require investing huge sums of money – is going to move forward.

As we work to transplant the i-mode business model overseas, we hope that other telecommunications providers will develop the confidence to adopt a 3G system themselves. Thus, one of the advantages of DoCoMo's international participation is that many telecoms in other countries will introduce the W-CDMA system, the 3G system DoCoMo supports, and offer international roaming. This means that our subscribers could use their own mobile phones when they are overseas. At present, because the standards used differ from region to region, it is not possible to use a Japanese mobile phone overseas. If more telecommunications providers adopt the same system as in Japan, the result will be of greater convenience for our subscribers.

6.8.3 Toward a Complex System on a Global Scale

International roaming is definitely a plus, but it is an advantage only for subscribers who travel overseas often. Actually, what I am hoping for, far more than international roaming, is that work on developing new handsets and new services will accelerate on a global scale. That would indeed, over time, be an advantage for our domestic subscribers.

The pace of development of mobile phones has undoubtedly picked up since the launch of i-mode. DoCoMo is receiving far more proposals for new features to be built into the next generation of handsets than it had earlier. And it is not only the manufacturers who are applying their creativity to mobile phone development. DoCoMo receives a steady stream of proposals for joint businesses from an extraordinary number of firms in a wide range of fields. Some of our new alliances have been initiated by us, but many result from proposals brought to us by other companies.

If DoCoMo sails into the international domain under the i-mode flag, then manufacturers and service providers from other countries will join our network of alliances. This will mean that new technical innovations and significant new cost reductions can occur on a global scale, benefiting subscribers in the form of enhanced convenience.

One partnership that may help start the ball rolling is our alliance with America Online (AOL). AOL is of course a pioneer in providing Internet services. Through alliances with it and other major players in the world of the Internet, we hope to stimulate the emergence of a functioning complex system on a global scale. Our next big objective is to expand our i-mode business into a business on a global scale. I want to develop the complex system that has formed in Japan into a global system. Global emergence, self-organization, and positive feedback: those are my goals.

Appendix 1

History of i-mode

Year	Month	Date	Event	Mobile phone models	Subscribers
1998	November	19	Press conference announcing i-mode		
1999	January	25	Press conference unveiling the i-mode commercial		
	February	1	Alliance with Pumatech, Inc., to develop applications for the corporate market		
		22	i-mode service launch	Digital Mova F501i	
		23	Testing of car navigation alliance started		
	March	16	Memorandum of intent over technical cooperation with Sun Microsystems		
		24		Digital Mova D501i, Digital Mova N501i	
	May	20		Digital Mova P501i	
	June	1	Forever Kyarappa daily cartoon character service launched		

(*continued overleaf*)

Year	Month	Date	Event	Mobile phone models	Subscribers
		16	Java-capable i-mode phone shown at JavaOne '99 Java developers' conference		
	July	12	Expanded functions for i-mode mail (user definable addresses, secret code function)		
		30	Expanded functions for i-mode mail (reception of 'Short Mail' service messages on i-mode phones)		
	August	8			1 million
	October		DoCoMo exhibited i-mode at Telecom '99, in Geneva		
		18			2 million
	December	3		Digital Mova D502i	
		23			3 million
2000	January	7		Digital Mova D502i	
	February	10		Digital Mova N502i	
		14			4 million
	March	4		Digital Mova P502i	
		15			5 million
		29	Invested in Playstation.co.jp Invested in Japan Net Bank, Limited		
	April	11		DoCoMo Nokia NM502i	
		15			6 million
		19	Invested in online payment service Payment First		
	May	26			7 million
	June	1	Establishment of D2 Communications joint venture with Dentsu Inc.	Digital Mova F502i, DoCoMo by Sony S0502i, Super Doccimo SH821i	

Year	Month	Date	Event	Mobile phone models	Subscribers
		20		Digital Mova F209i, Digital Mova N209i, Digital Mova P209i	
		23	Added English menu and content		
		24			8 million
		30	Lent English-version i-mode phones to members of the press at the G8 Kyushu–Okinawa Summit		
	July	7		Digital Mova D209i	
		14		Super Doccimo N821i	
		15			9 million
	August	1	Memorandum of intent to cooperate with Sony Computer Entertainment in technology development		
		5			10 million
		19		Digital Mova P209iS	
		28			11 million
	September	1		Digital Mova N209it	
		6	Established i-mode center in Yokohama, moved to distributed processing		
		8		Super Doccimo P821i	
		18			12 million
		19	Established Japan Net Bank, Limited		
		27	Alliance with AOL for joint development of new Internet services		
		29	Memorandum of intent with KPN Mobile concerning mobile Internet operations		

(*continued overleaf*)

Year	Month	Date	Event	Mobile phone models	Subscribers
	October	5	Alliance of DoCoMo, Lawson, Matsushita Electric Industrial, and Mitsubishi Corporation to create i-Convenience, Inc.		
		8			13 million
		20		Digital Mova R209iS	
		27	Established DoCoMo.com as a consulting company specializing in i-mode content		
		31			14 million
2001	Spring		Scheduled to launch G3 service Scheduled to launch service in alliance with PlayStation		

Appendix 2

Proposal for the Basic i-mode Concept

The following is the proposal for the basic i-mode concept that I submitted in September 1997. I have not altered it in any way.

Confidential
September 12, 1997

Memorandum

Concerning the Gateway Project Plan

1. Basic Gateway Project Concepts

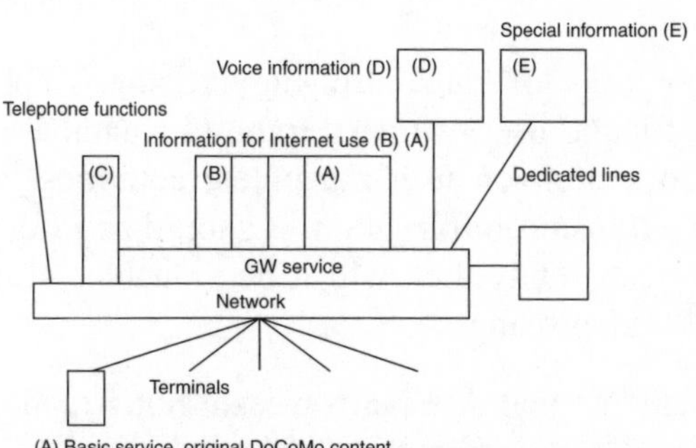

(A) Basic service, original DoCoMo content.
(B) Services provided by content providers. By using the Gateway service, they can deliver the necessary information or services to the people who need them, an objective difficult to achieve on the Internet.
(C) Services that are not dependent on Gateway servers and databases, for example, viewing Internet content as is. That is, extensions of telephone functions.
(D) Transmission of information by voice or mixed voice and data content. Could include telemarketing.
(E) Linkage with data that must be kept secure, for example corporate closed user groups, databases, settlement data.

What the Gateway service seeks to do is, to put it clumsily, 'provide a new infrastructure for daily life'. That is, the mobile phone is to be an essential part of life. Mobile phones become in effect integral parts of their users' bodies, enabling them to complete actions that, a century ago, would have required physical transmission in person or by a letter – making contact with another person, searching for information, gathering information, purchasing goods and services. Most of these functions are already available on the Internet, but the Internet has two disadvantages. Access requires a computer, and the content available is too rich; it requires considerable effort to find the right content or the right information one wants. The Gateway project will give priority to the portability of mobile handsets and leave the acquisition of graphics and detailed information to computer-based services. We will focus on establishing the mobile terminals and Gateway services as essential infrastructure for daily life by providing information and services in ways that are focused on everyday utilities.

'DoCoMo is always with you from cradle to grave.'

'Don't forget it, even when you're running away.'

'Deathbed request: Let me do a full reset.'

Using the Internet is only one possible interface between content providers and users. We need not be fixated on providing the same service. We should proceed by using Internet content as our base while considering voice, dedicated line connections, and other possibilities as the nature of the data dictates.

Databases are not tools for transmitting advertising. To provide information that suits a particular user, we need to build a database for organizing and analyzing information on user attributes, activities, and purchasing histories. When we transmit information, it should be information that fits the recipient's interests. Only then will it be valuable information to him or her, even if it is advertising.

'Advertisements that Are Narrowcast, Not Broadcast, Gain Information Value'

2. Gateway Services Seen from Various Perspectives

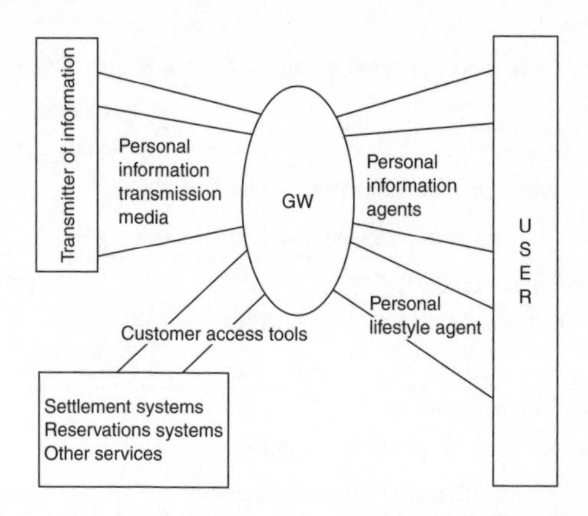

Gateway services will be capable of establishing multimedia linkages between content and user. They will make it possible to communicate the content provider's information or service to precisely the users that need them.

'The ultimate in personal media.'

To the subscriber, Gateway services will turn a mobile phone into a 'personal agent', that is, a personal concierge that transmits – from all the various kinds of information available – information tailored to one's own attributes. Of course, the user can also deliberately seek information.

In addition to providing a gateway to Internet content, we will consider extensions of Gateway services such as personal tools, with imbedded individual ID functions and hybrids of IC card functions. These are what we call 'Personal lifestyle agent' functions.

Basically, our aim is to stimulate use of Gateway services for all commercial transactions and information distribution using mobile phones by making it advantageous for information providers and content providers. (Locking in potential customers who use push features, targeted messages using databases, billing and collection by DoCoMo, etc.)

3. Schedule

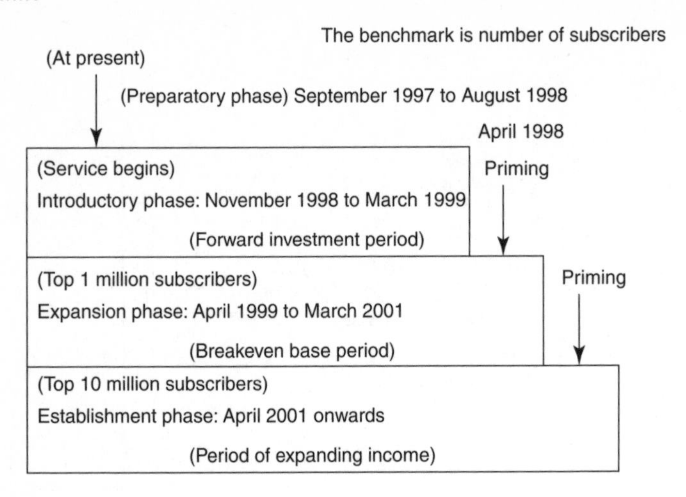

1. Introductory phase

During the introductory phase, it will be crucial to show both subscribers and information providers specific examples. To achieve that objective, we should consider actively developing alliances with content providers, to provide content of increased utility and convenience to subscribers. In this phase, the DoCoMo Content Team will lead a main group of 10 to 20 content providers and also invest in original content. We will have a clear policy on content selection in this phase. The subscriber fees for accessing content should be as low as possible, while push-type advertisements for content will be free of charge. Our goals are to energize the information flow on the Gateway service and expand our subscriber base.

Candidates for inclusion in this phase include content in the following categories:

Everyday utility, with push-type linkage:

- Travel: Company H, Company A, others (provide registration information)
- Airlines: Airline J, Airline A (alliance with FFP)
- Banks: Bank C or Bank S (banking transactions)
- Real estate (rentals): Company A (linkage between providing information and actual transactions)
- CD sales: Company T (sales, trial listening)

- Book sales: Company K (sales)
- Others.

Lifestyle information type, basically pull type:

- Weather: Company W, Company T
- News updates: newspapers
- Information retrieval
- Employment information: Company R
- Others.

Advertising and announcement type, both push and pull:

- An advertising medium providing information on special offers, combined with promotions
- Providing advertising information in a format that is advantageous to the subscriber, combined with telemarketing: Company N, Company B.

Local information – maps and other local reference, push-type linkage:

- Harajuku shopping area
- Haneda Airport, Tokyo Station
- Others.

Community type:

- Restaurant rankings, based on subscriber votes
- Forums, chat rooms, e-mail, instant messaging
- Others.

2. Expansion phase

During this phase, while building on the success of the first phase, we will readily accept content other than that selected by DoCoMo. For this phase to work, we must exercise care in establishing specifications, including use of HTML, and other rules from the start. Our aim is to make it relatively easy for others to provide content voluntarily, without input from DoCoMo.

When we have attained critical mass, with over one million subscribers, DoCoMo will no longer have to encourage the development of content. Potential content providers will be aware of the value of using Gateway as a medium. That means that, just as with the Internet, we will be awash with content, which will be too much for both subscribers and content providers. At this stage, Gateway must concentrate on expanding agent functions using databases.

While how to develop useful content is a critical issue in the launch phase, during the expansion phase, the issue shifts to the need to organize the information and provide navigation tools. Here for the first time the effort put into the Gateway database will begin to pay off.

Content from some virtual communities may also appear in this phase. With improvements in mobile phone displays and input functions, it also will be possible to provide richer content than in the launch phase. These developments should be actively encouraged, since they provide a testing ground for the third phase.

At this point, the Gateway business model should begin functioning. That is, we will charge for use of the database and push transmission of content. Of course, with use of the database, push transmission will be properly targeted and not perceived as intrusive by subscribers.

Also, we will begin to work actively from this stage on making the mobile phone itself more multifunctional (with IC cards, etc.).

3. Establishment phase

When we have reached this phase, life without a mobile phone will be impossible for subscribers (or so they will feel). What we should focus on in this phase is expanding personal agent functions.

Having reached this stage, we expect spontaneous emergence of content of all sorts, as in the Internet today. DoCoMo will not have to lead content development.

Subscribers, however, will be fed up with too much content. Thus, we will use our database to expand personal agent functions. Depending on the nature of the service, we may charge subscribers.

We will continue to charge content providers as in the second phase, so that the business model has considerable earnings potential. A direct marketing medium reaching 10 million persons is unprecedented, worldwide.

Because the mobile phone and network specifications will be adapted to handle Internet content, mobile phones will always be superior to computers.

Appendix 3

The i-mode Menu List

Top Menu (as of November 6, 2000)
  Top menu
  By region
    News, weather, info
    Mobile banking
    Credit cards, stock brokerages, insurance
    Travel, transportation, maps
    Shopping, lifestyle
    Gourmet interests, recipes
    Local information, government offices
    Dictionaries, handy tools
    e-mail
News, weather, info
  News, weather, info
    Share prices, investor information
    Regional newspapers
    Overseas media
    Sports, outdoor news
    TV, FM
Mobile banking
  Mobile banking
    Nationwide banks
    Regional banks
    Credit associations, nationwide
    Credit cooperatives, nationwide

        - Agricultural cooperative associations
        - Banks, by name
        - Credit associations, by name
        - Agricultural cooperative associations, by name
        - Credit cooperatives, by name
- Banks, by name
- Regional banks
    - Regional banks
        - Full service
        - Balance inquiry
- Credit associations, by name
- Agricultural cooperative associations, by name
- Entertainment 1
- Ringtone melodies, images
    - Ringtone melodies, karaoke
    - Characters
    - Visuals
    - Games, fortune-telling
    - Games
    - Fortune-telling, diagnosis
- Entertainment 2
    - Entertainment
        - Sports, outdoors
        - Music and movie info
        - Prizes, lotteries, horseracing
        - TV, FM
        - Magazines, variety
        - Show business
        - Local stations, etc.
- Games, search by type
    - Games
        - Search screen top
        - Role-play games
        - Simulations
        - Sports games
        - Educational games
        - Communication games
        - e-mail games
        - Table games

Quizzes
Novelistic games
Variety games
Ringtone melodies, characters
Show business celebrities, by name
Dictionaries, handy tools
Dictionaries, handy tools
i-mode handy numbers
Delivery status inquiries
Mobile phone manufacturers
English menu
English menu

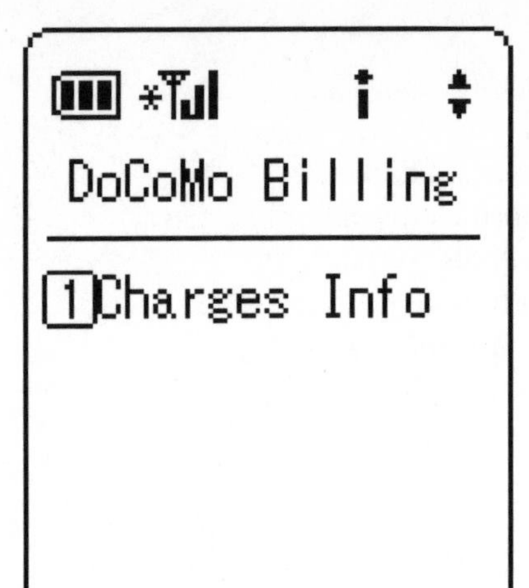
DoCoMo Billing
1 Charges Info

BK Mobile Banking
1 Citibank
¥ Trading
2 TMTDW

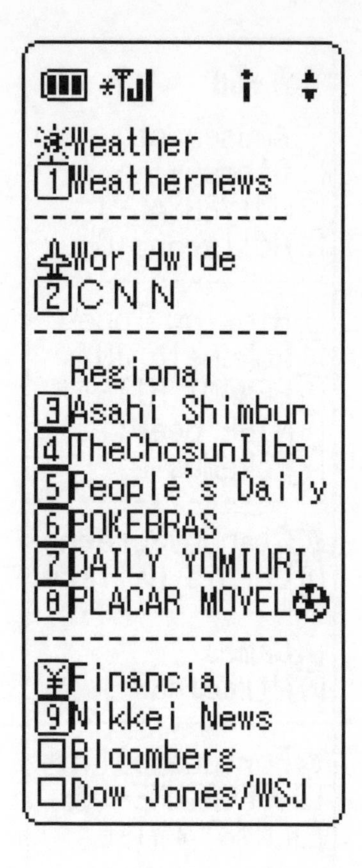
Weather
1 Weathernews
Worldwide
2 CNN
Regional
3 Asahi Shimbun
4 TheChosunIlbo
5 People's Daily
6 POKEBRAS
7 DAILY YOMIURI
8 PLACAR MOVEL
Financial
9 Nikkei News
Bloomberg
Dow Jones/WSJ

Airlines
1 Northwest Air
2 Cathay Pacific

Amusement
1 Disney-i
2 UNIVERSAL-J
3 Hollywood Ch.
Ringing Tones
4 Pokemelo JOY
5 BEMANI HITS!
6 VIBE beep!!
7 PokeMusic
StandbyScreens
8 Corbis Gallery
Games
9 MiracleGP
FortuneTelling
★FORTUNE-i★
LUV♂♀Wise
Communication
imaHima!

1 News/Info
2 Banking/Trade
3 Travel
4 Entertainment
5 Lifestyle
6 DoCoMo Billing
日本語

Further Reading

For more information on complex systems, I recommend the following books.

Casti, J. L., *Complexification*, HarperCollins, New York, 1994.
Minsky, M. L., *The Society of Mind*, Simon & Schuster, New York, 1986.
Thaler, R. H., *The Winner's Curse: Paradoxes and Anomalies of Economic Life*, Free Press, New York, 1992.
Waldrop, M. M., *Complexity: The Emerging Science at the Edge of Order and Chaos*, Simon & Schuster, New York, 1992.

Afterword to the Japanese Edition

What I wanted to do with i-mode

In 1993 – that was before Internet access was widespread – I enrolled on the MBA program at the Wharton School of the University of Pennsylvania. Upon arriving in the United States, I promptly purchased a Macintosh and signed up for America Online (AOL), having heard about its rapid growth. I had some experience accessing what was then called an online service for personal computers – CompuServe – from Japan via Nifty-Serve, a similar Japanese service, but only for doing research. I had not made use of CompuServe in my private life.

I was amazed. AOL's simple setup and the graphical user interface (GUI) were good, but what really thrilled me were the services it offered. Even in 1993, I could use easySABRE, operated by a subsidiary of American Airlines, to search for seats and fares, including discount tickets, on airline routes throughout the world, and then purchase the tickets online. I could also make hotel and car rental reservations online. American Express let me check my online credit card usage details and how much my next month's bill would be. Online direct sales were also booming. In fact, what we now call e-commerce was thriving in the United States before Internet access became widespread.

AOL also offered extensive flash news and information search services. It was great fun to search for articles from past issues of *Newsweek* or check CNN news. As I used credit card, travel, and other services closely tied to my personal life, I became increasingly aware of their convenience. Why, I could work out all the details of the itinerary for a trip at home, at any time of the day or night, without contacting a travel agency. In mastering the system, I learned about such things as season passes and weekend discounts and felt I was getting as knowledgeable as a travel professional.

I had another surprise when school started and I learned the details of the curriculum. Wharton was offering a course on the impact of the Internet on business! I knew a little about the Internet, but had not reached the point of considering its potential impact on real businesses. I promptly signed up for that course, and as my studies continued, became increasingly aware of what an amazing development the Internet could become. I realized that what I was experiencing in being able to buy airline tickets on AOL could, thanks to the Internet, extend to all industries.

'Life is getting more convenient and more full of opportunities!'

It was true. By the time I graduated and returned to Japan, in 1995, high-tech start-up companies were popping up and the Internet had become a household name in the United States. All sorts of industries were starting to offer services via the Internet. For me, personally, life was getting more and more convenient.

'Why isn't this possible in Japan?' That was the question that I could not help asking after I returned to Japan. I could use AOL to check seat availability by class of service and fares, including discounts, for JAL flights from Narita to New York. Why was there no such domestic service? It would make such a difference in Japan too, making life more convenient and full of opportunities.

While studying in the United States, I also encountered complex systems theory for the first time. When we were studying the economics of cities, complex systems theory was introduced to explain how urban functions accumulate.

Until then, I had found the social sciences I studied unrealistic and unattractive, because their approach focused on isolating factors and analyzing them. To carry out that analysis, social scientists built models based on entirely unrealistic assumptions. The complex systems approach, with its notion that the world is complex and results keep changing depending on how the component elements relate to each other, fascinated me.

The idea that complex systems theory would apply to the Internet world had occurred to me before I joined the i-mode project, when I was still vice president of Hypernet, a Net start-up company.

In all the social sciences, there is a huge difference between intellectual understanding, reality, and practice. Studying can lead to intellectual understanding, but putting that into practice is another matter.

Implementing ideas requires a grasp on reality. That was brought home to me most acutely while I was at Hypernet.

Returning to Japan full of confidence that I had soaked up all there was to know about the most advanced aspects of the Internet, I met the President of Hypernet and, at his invitation, joined that company. The launch of this new business, which offered free Internet access supported by advertising sales, went smoothly, but we were not able to expand it as much as we had planned. The idea was not bad. It was even socially significant. If Hypernet had developed as planned, it would have made Japan a better place to live. We had excellent employees and they worked incredibly hard. We were sure that we were not mistaken in what we were doing. But it did not work. Why not?

For some reason, I happened to remember complex systems theory. Seen within the larger societal framework, a company of just 80 people trying their best to do something cannot have a large impact – no matter how right what they are trying to do is. Dazzled by the Internet and start-up company boom, we had fallen into the illusion that if we had a few good people, the market would follow us. A few people, however, cannot have a transformative effect on society. It takes many people starting new things, little by little, to effect a major transformation. The number of participants matters. Social change requires critical mass.

The reason Japan did not have leading-edge Internet services as in the United States was, basically, that too few people would use them. With few potential users, why would an airline, for example, develop an online service? And with no convenient services to attract them, the number of users could not grow. Japan was in a negative feedback cycle, Internet-wise.

I had just begun to awaken to that reality while slaving morning to night to make our business go when I had a phone call from Mari Matsunaga, author of the *i-mode Incident*.

'I'm changing jobs.'

'What, the editor of a job-change magazine is doing it herself?'

'Yes, it's true, and on top of that, I'm moving to DoCoMo. I'm going to be part of the old phone monopoly!'

'What will you be doing?'

'Developing some sort of service to exchange data by mobile phone.'

I was struck dumb. That was it! Mobile phones have a vast number of users. They even outnumber personal computer owners in the United States. This could be big. It could be the way to bring the convenience and opportunities I had hoped for to Japan.

This book explains the strategies behind the explosive growth of i-mode. At first glance, it may appear that we had everything taped, but we were constantly making adjustments as circumstances dictated, without, of course, changing our basic thinking. In a complex systems world, nothing is absolute. Everything is relative, and the accidental is decisively important.

Of the many accidents that contributed to the success of i-mode, the most important was the group of people who chanced to come together in the i-mode project: Koji Ohboshi, now Chairman of DoCoMo (then President); Keiichi Enoki (now a DoCoMo Director), assigned by Ohboshi to have overall responsibility for i-mode; Enoki's recruit, Mari Matsumoto, godparent of i-mode; and the many others who have worked on i-mode. Had one of them not been part of the project, the outcome would have been very different.

That is true not only of the team within DoCoMo but of our partners as well. If Sumitomo Bank had not decided to participate, if Bandai had not launched 'Forever Kyarappa', if a manufacturer had not suggested downloading MIDI-format ringtones . . . .

What I have contributed is the basic strategy, as described in this volume. More important than the strategy itself, of course, is applying it, extending and interpreting it, and making the necessary changes in response to changing circumstances. That, as complex systems theory makes clear, is far more important – and far more difficult. For that reason, I have the highest admiration for every NTT DoCoMo employee, including the current top management, from President Keiji Tachikawa on down, and our brilliant engineering team, and I am honored to be numbered among those employees.

Finally, I would like to note that complex systems were at work in the production of this book as well. Emergence and self-organization occurred in almost every aspect – title, structure, table of contents, index, binding – so that the book is far more than what I had initially imagined. I am grateful to the people at Nikkei BP, including Hisashi Arai (Senior Executive Managing Director, Nikkei BP Planning) and Yukiko Ichimoto for making this volume possible.

Takeshi Natsuno
On an ANA flight to New York
October 2000

Index